Small Ring Compounds in Organic Synthesis II

Synthesis and Synthetic Applications of 1-Metallo-1-Selenocyclopropanes and -cyclobutanes and Related 1-Metallo-1-silylcyclopropanes

A. Krief

Cyclopropenes and Methylenecyclopropanes as Multifunctional Reagents in Transition Metal Catalyzed Reactions

P. Binger and H. M. Büch

AKADEMIE-VERLAG BERLIN

Topics in Current Chemistry · Vol. 135

Small Ring Compounds in Organic Synthesis II

Small Ring Compounds in Organic Synthesis II

With 5 Figures and 11 Tables

AKADEMIE-VERLAG BERLIN
1987

Die Originalausgabe erscheint im Springer-Verlag Berlin · Heidelberg · New York · Tokyo
in der Schriftenreihe „Topics in Current Chemistry“,
Volume 135

Vertrieb ausschließlich für die DDR und die sozialistischen Länder
Akademie-Verlag Berlin

ISBN 3-540-16662-9 Springer-Verlag Berlin Berlin · Heidelberg · New York · Tokyo
ISBN 0-387-16662-9 Springer-Verlag New York · Heidelberg · Berlin · Tokyo

ISBN 3-05-500284-9 Akademie-Verlag Berlin

Erschienen im Akademie-Verlag Berlin, DDR-1086 Berlin, Leipziger Straße 3—4
Lizenznummer: 202 · 100/467/87
Printed in the German Democratic Republic
Gesamtherstellung: VEB Druckerei „Thomas Müntzer“, 5820 Bad Langensalza
LSV 1275
Bestellnummer: 763 702 5 (3072/135)
12800

Editorial Board

Table of Contents

Table of Contents of Volume 133

Small Ring Compounds in Organic Synthesis I

Synthesis and Synthetic Applications of 1-Metallo-1-Selenocyclopropanes and -cyclobutanes and Related 1-Metallo-1-silylcyclopropanes

Alain Krief

Facultés Universitaires de Namur, Départment de Chimie Rue de Bruxelles 61, 5000 Namur, Belgium

Table of Contents

Among the α-heterosubstituted cyclopropylmetals α-selenocyclopropyllithiums represent some of the most valuable synthetic intermediates. They are quantitatively prepared from selenoacetals of cyclopropanones and butyllithiums, are thermally stable at $\sim -78°$ for several hours and are particularly nucleophilic especially towards carbonyl compounds. The cyclopropyl derivatives containing a selenenyl moiety have been transformed to selenium free derivatives such as alkylidene cyclopropanes, vinyl cyclopropanes, allylidene cyclopropanes, cyclobutanones and α-silyl cyclopropyllithiums. The latter compounds have been used as starting material for the synthesis of alkylidene cyclopropanes and cyclopentenyl derivatives. α-Seleno cyclobutyllithiums, which are available in two steps from cyclobutanones, also permit the synthesis of various selenium free homologues such as alkylidene cyclobutanes, vinyl cyclobutanes, oxaspirohexanes and cyclopentanones.

1 Background

The presence of a selenenyl moiety in organic molecules confers on them unique properties [1–12]. The selenium atom in selenides is particularly nucleophilic towards, for example, alkyl halides and halogens [1,2] (Scheme 1); it is oxidizable leading

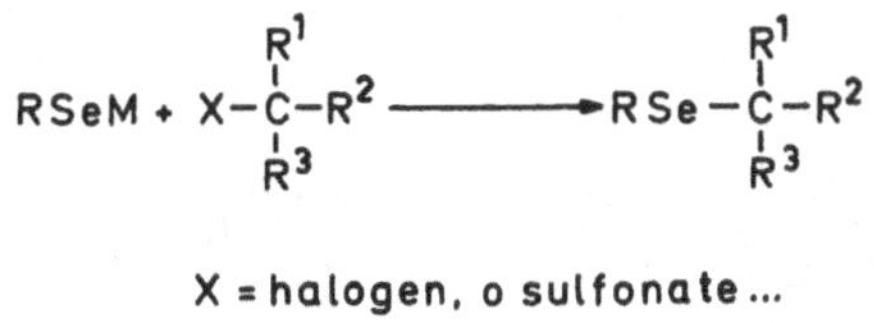

Scheme 1

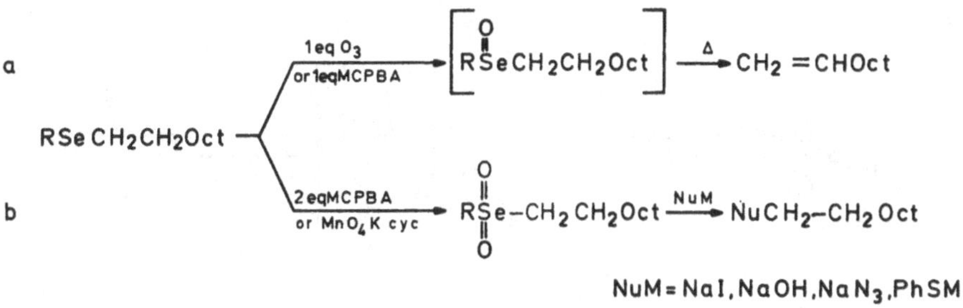

Scheme 2

selectively to selenoxides [3–9,11,12] under mild conditions (Scheme 2a) or to selenones with excess of oxidant [13,14] and under more drastic conditions (Scheme 2b). The selenium atom is also electrophilic: selenides react with alkyllithiums and lead to novel selenides and novel organometallics by cleavage of the original C—Se bond (Scheme 3) [7–9,12].

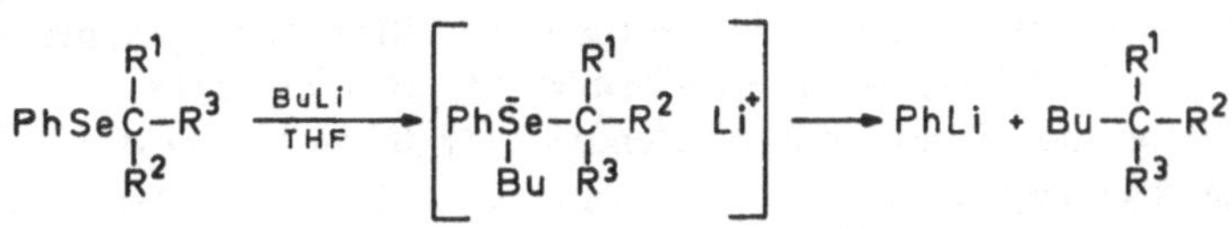

Scheme 3

Finally the selenenyl moiety is perfectly able to stabilize a carbanion [7–9,12] (Scheme 4a) or a carbenium ion [9,10,12,15] (Scheme 4b). The selenenyl moiety in

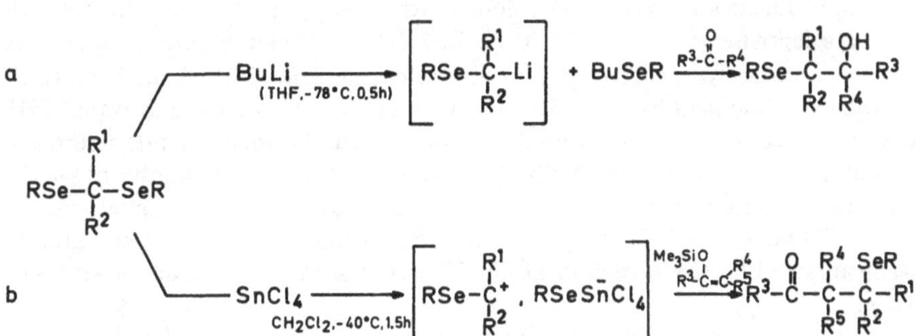

Scheme 4

selenides can be removed by a large variety of reagents. Substitution reactions are observed in several instances. For example, selenides are reduced to alkanes (Raney-Ni, Li/NH_3, $HSnR_3$) [16–18] (Scheme 5) or transformed into alkyl halides on

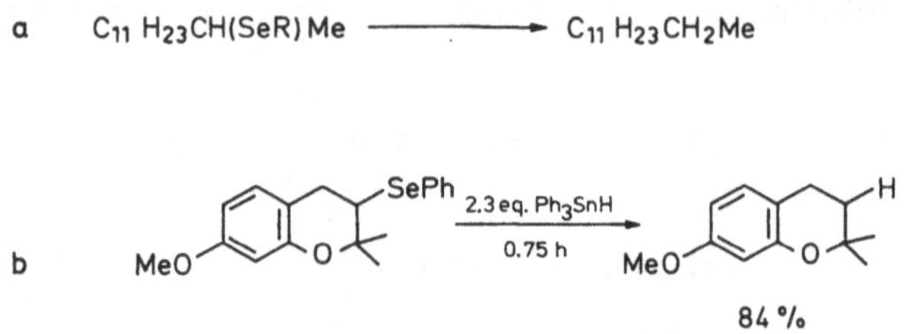

Scheme5

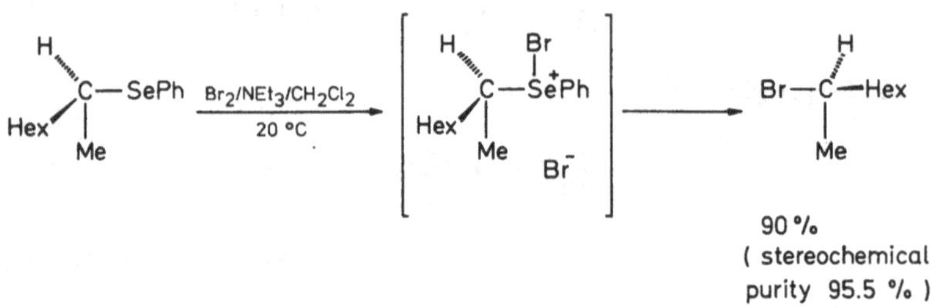

Scheme 6

direct reaction with bromine [19,20] or methyl iodide [7] (Scheme 6). The selenenyl moiety can be transformed to a better leaving group, such as a seleninyl or a selenonyl group. The oxidation to selenoxide is usually very easy [3–9,11,12] and takes place even at low temperatures (—78 °C) when ozone is used. Formation of selenone is rather difficult due to the competition of the selenoxide elimination reaction [6,13]. A few reports deal with the substitution of the selenenyl moiety: for example, selenides have been transformed to alkyl chlorides and bromides on reaction [21] of the corresponding selenoxides with hydrochloric or hydrobromic acids. The selenonyl moiety in selenones is a much better leaving group [14] (even better than the iodide ion [14]) which posseses a high propensity to be substituted rather than to be eliminated. Thus selenides produce, through the selenones [13,14], alkyl iodides (NaI, P_2I_4), alkyl bromides or chlorides (MgX_2, RMgX), alkyl sulfides (PhSM), alkyl azides (N_3M) and alcohols (KOH) (Scheme 2b).

Elimination reactions leading to olefins are usually performed on the corresponding selenoxides [3–9,11,12] (Scheme 2a). These are often unstable and decompose at room temperature to olefins and selenenic acid (further oxidized to the more stable seleninic acid by excess of oxidant). Hydrogen peroxide in water-THF, ozone and further treatment with an amine or tert-butyl hydroperoxide without or with alumina proved to be the method of choice for such a synthesis of olefins. The reaction is reminiscent of the one already described with aminoxides or sulfoxides [22] and occurs via a syn elimination of the seleninyl moiety and the hydrogen attached to the β-carbon atom. However it takes place under smoother conditions.

Olefins can also be produced [23] by reaction of selenonium salts with bases.

Again a syn elimination reaction, involving now the carbanion present in the ylide moiety, has been invoked (Scheme 7).

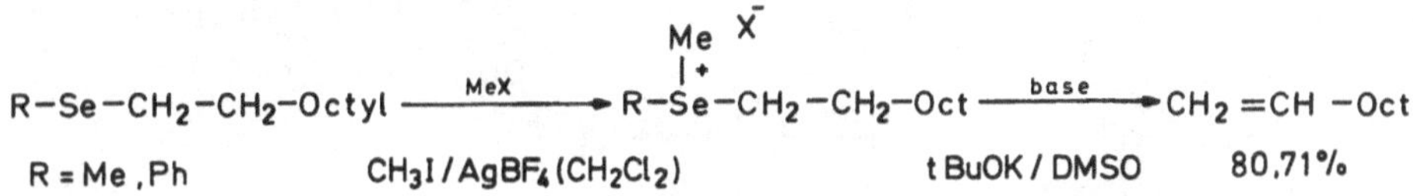

Scheme 7

Among the functionalized selenides, β-hydroxy-alkyl-selenides [3–9,11,12] and allylselenides [3,24–41] are those which possess a typical reactivity.

In some way β-hydroxy selenides resemble pinacols in their reactivity (Scheme 8).

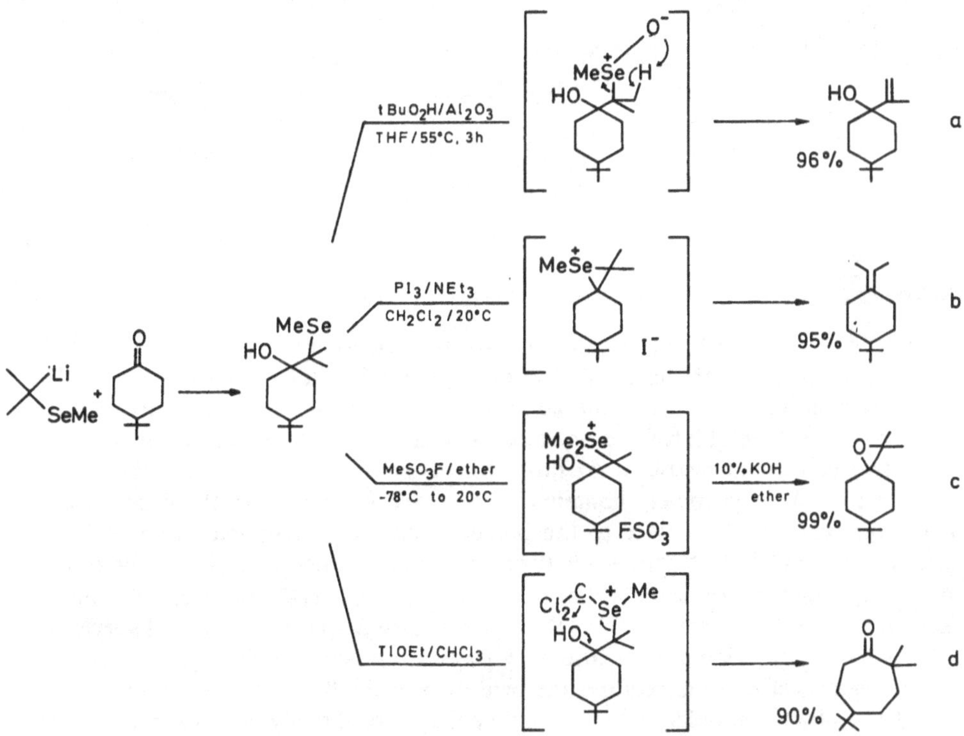

Scheme 8 A

The presence of the soft selenium atom and the hard oxygen however, make, the reaction of β-hydroxy selenides site selective. These have in fact been transformed selectively to vinylselenides [7] or olefins [4–9,11,12], by selective activation of the hydroxy group, inter alia, with thionyl chloride alone or with thionyl, mesyl and phosphoryl chloride, trifluoracetic anhydride, phosphorus triiodide or diphosphorus tetraiodide in the presence of triethyl amine (Schemes 8Ab; 8Bb). The formation of olefins from β-hydroxyselenides is regio- and stereoselective and occurs by formal removal of the hydroxyl and selenenyl moiety in an anti fashion.

Selective activation of the selenenyl moiety of β-hydroxy selenides has been achieved with methyl iodide, dimethyl sulfate or methyl fluorosulfonate. The selenonium salts produced have been transformed to epoxides [3–9, 11, 12, 35] on treatment with a base (aq. KOH/ether, and tBuOK/DMSO, inter alia) (Schemes 8 Ac;

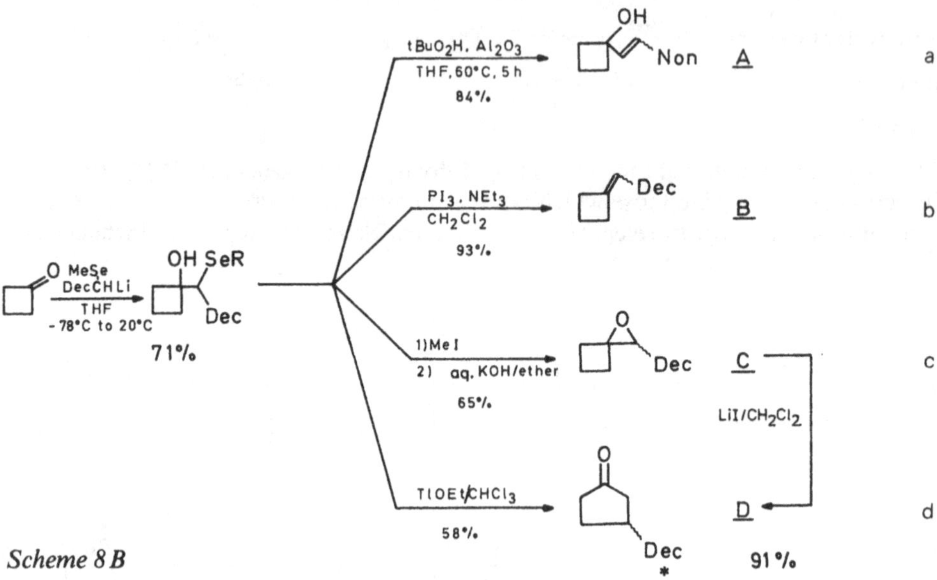

Scheme 8 B

8 Bc). The reaction is highly stereoselective, the selenonium salt being substituted with a net inversion of the configuration at the substituted carbon atom [3]. The same reaction has also been achieved [42] in one pot from β-hydroxy selenides and thallium ethoxide in chloroform or aqueous potassium hydroxide in chloroform. In these two cases it is restricted to those β-hydroxy selenides in which the carbon bearing the selenenyl moiety contains at least one hydrogen. In the other cases a rearrangement, which is close to the pinacolic rearrangement, but completely site selective, occurs [8, 12, 43, 44] and leads to aldehydes or ketones which retain the oxygen originally present on the same carbon atom (scheme 8 Ad; 8 Bd). Both transformations have been found to proceed [8, 12, 44] through dichloro carbene (or a related species).

β-hydroxy-alkyl-selenides are also very powerful precursors of allyl alcohols [3–9, 11, 12]. The transformation requires the oxidation of the β-hydroxy-alkyl-selenides to β-hydroxy-alkyl-selenoxides which usually collapse to the allyl alcohol below 70 °C and often at room temperature. Hydrogen peroxide supported on alumina in THF are, among the conditions reported, the ones which can be recommended (Schemes 8 Aa, B Ba).

Allylic selenides, with a substitution which is different at the terminal carbon-carbon double bond and at the carbon bearing the selenenyl moiety, are often unstable and rearrange [25] to the thermodynamically more stable allylic selenides, which in fact possess the more highly substituted carbon-carbon double bond. The isomerisation of the phenylseleno derivatives is efficiently achieved in sunlight or with fluorescent bulb in the laboratory after less than 1 hr [25]. Methylseleno analogues are very sensitive to traces of acid and rearrange [25] almost instantaneously, even on buffered SiO_2 TLC plates.

Oxidation of these allylic selenides with ozone, hydrogen peroxide, and sodium periodate does not lead to the expected selenoxides but produces, in almost quantitative yield, allyl alcohols resulting from a selenoxide-seleninate rearrangement [24, 26, 29, 41, 45]. Similarly, allylamines are formed [38] when allyl selenides are reacted with chloramine T.

With these interesting types of reactivity of selenides and functionalized selenides, it is important to show that these compounds can be rapidly prepared from readily available starting materials. At least two types of methods are available for such purposes and involve the attack of a selenolate [1-4] or of an α-seleno alkylmetal [4-9, 11, 12], on an electrophilic carbon atom. This last reaction is particularly interesting since a new carbon-carbon double bond is formed in the process.

Little was known about the synthesis and the reactivity of α-selenoalkylmetals prior to our work. It has now been clearly established that any α-selenoalkylmetal with a carbanionic center bearing a hydrogen and/or an alkyl group cannot be prepared by metallation of the corresponding selenides [7]. This can be rationalized in that the selenenyl moiety does not sufficiently stabilize a carbanion and consequently a base such as a dialkylamide is not strong enough to metallate a selenide (or a sulfide), and alkyllithiums, which are strong enough to perform the hydrogen-metal exchange in sulfides possessing a similar acidity, cleave instead the carbon-selenium bond in selenides.

Such a propensity of the carbon-selenium bond to be transformed into a carbon-lithium bond on reaction with butyllithiums has in fact been used successfully for the synthesis of various α-selenoalkylmetals from phenyl and methyl selenoacetals. It has inter alias been used for the synthesis of those α-selenoalkylmetals which bear two alkyl groups on the carbanionic center and which are expected to be the less stabilized ones [3-9, 11, 12]. It also permits the selective synthesis of α-lithioselenoacetals from selenoorthoesters [8, 9, 12].

Although unable to metallate selenides, dialkyl amides are sufficiently strong to metallate phenylselenoacetals [39, 46-51] as well as methyl [48, 52] and phenyl [46, 47, 52] selenoorthoesters. They are also able to metallate selenoxides [4-9, 11, 53-55] and selenones [14]. Finally selenoacetals are readily available [4, 7, 11, 12, 56] from carbonyl compounds and selenols in the presence of a Lewis acid and selenoorthoesters have been prepared from orthoesters, selenols, and boron trifluoride etherate [47, 48, 52].

2 Syntheses of 1-Functionalized-1-Metallo Small Ring Compounds

A few years ago we became interested in adapting the synthetic methods mentioned above to the cyclobutyl and cyclopropyl derivatives. The strain present in such compounds must be taken in to account. For example, cyclopropanone is not a suitable starting material for the synthesis of the corresponding selenoacetal due to its instability, and alkylidene cyclopropanes are more diffficult to prepare than other olefins, due to the strain present. The methods listed in the first section proved in several instances inefficient, and a search for new reagents was often required to achieve the goal. The strategy we will discuss involves: a) the synthesis of α-metallocyclopropyl derivatives bearing a selenenyl, a seleninyl, or a selenonyl moiety;

b) their reaction with an electrophilic carbon atom; and

c) the removal of the selenyl moiety of the resulting compound in order to prepare selenium-free derivatives [7,8,12].

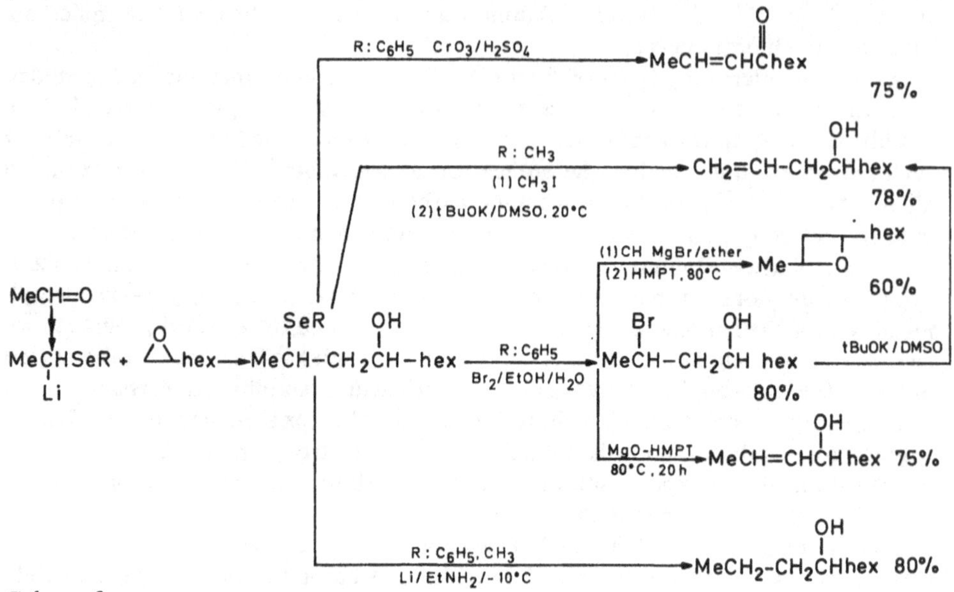

Scheme 9

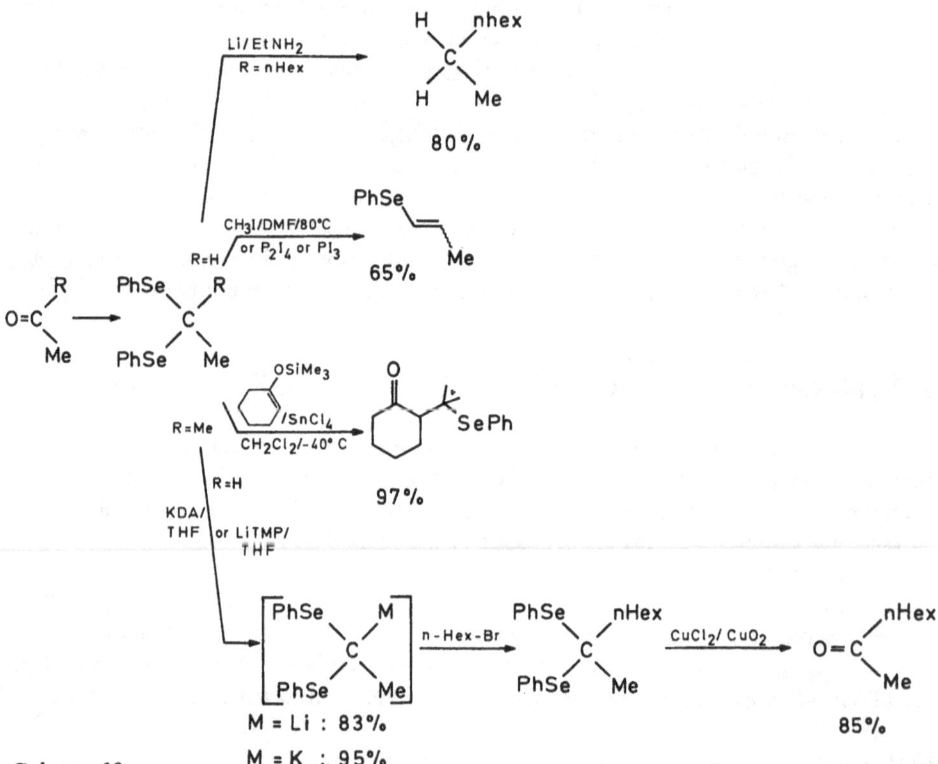

Scheme 10

Scheme 8, 9 and 10 disclose specific examples of such a strategy applied to α-selenoalkyllithiums which do not belong to the cyclopropyl or the cyclobutyl series.

Cyclobutyl compounds were found [57] to have reactivities closely related to the ones already disclosed for open chain and other cyclic derivatives [7,8,12]. Moreover, in several instances, compounds possessing a cyclobutane ring have been prepared from α-selenoalkyllithiums and cyclobutanones (Compare Scheme 8B to 8A) using the strategy already presented. This is not, however, the case of cyclopropane analogs due to the unavailability of cyclopropanones.

2.1 Syntheses of Functionalized (1-Seleno-, 1 Silyl-, 1-Vinyl-)Cyclopro-Phenyllithium

α-Selenocyclopropyllithiums and α-silylcyclopropyllithiums belong to the well-known family of α-heterosubstituted cyclopropyl metals [7]. The presence of the cyclopropyl ring enhances the stability of the carbanion and therefore favors its generation more than that of the corresponding heterosubstituted organometallic part of a larger ring or one bearing two alkyl groups on the carbanionic center.

Several α-heterosubstituted cyclopropyl metals are known. They have been prepared by:

a) hydrogen-metal exchange from the corresponding carbon acid *1* (Scheme 11a)

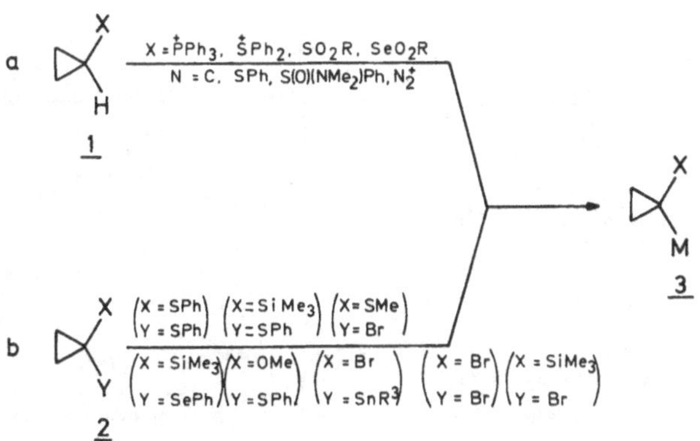

a)	x	Ref.	b)	x	y	Ref.
	$^+PPh_3$	58–60)		SPh	SPh	79, 80)
	$^+SPh_2$	61, 62)		$SiMe_3$	SPh	81a, 82)
	SO_2R	65, 66)		SMe	Br	71)
	SeO_2R	67,		$SiMe_3$	SePh	77,
	N≡C	68, 69)		OMe	SPh	84, 85)
	SPh	70)		Br	SnR_3	83)
	S(O) (NMe)$_2$Ph	63)		Br	Br	71, 73)
	N_2	64)		$SiMe_3$	Br	77, 78)

Scheme 11

b) halogen-metal exchange (Scheme 11 b) [SMe [71]), Br [71-73]), Cl [74]), $SiMe_3$ [75-78])]
c) heteroatom-metal exchange (Scheme 11 b) [SPh [79, 80]), $SiMe_3$ [77, 81, 82]), Br [83]), OCH_3 [84, 85]), SeR [86, 87])] which implies sulfur-lithium [79-82, 84, 85]), selenium-lithium [77, 86, 87]), or tin-lithium [83]) exchange.

Hydrogen-metal exchange is most frequently used, because the compounds to be metallated are easily synthesized. However, it lacks generality and often applies exclusively to the parent compounds [7]) or to those derivatives which possess another group (such as a vinyl, phenyl, or carbonyl group) able to stabilize the carbanionic center.

2.1.1 Attempted Syntheses Using Hydrogen-Metal Exchange

It was expected that the extra stabilization provided by the cyclopropyl group [88-93]) would be sufficient to permit the metallation [35, 39]) of cyclopropyl selenides [35, 39, 94, 95]) or of cyclopropyl silanes [96-98]), but that proved not to be the case. The phenylseleno and methylseleno cyclopropanes required for this study were prepared by the routes outlined in Scheme 12, which involve:

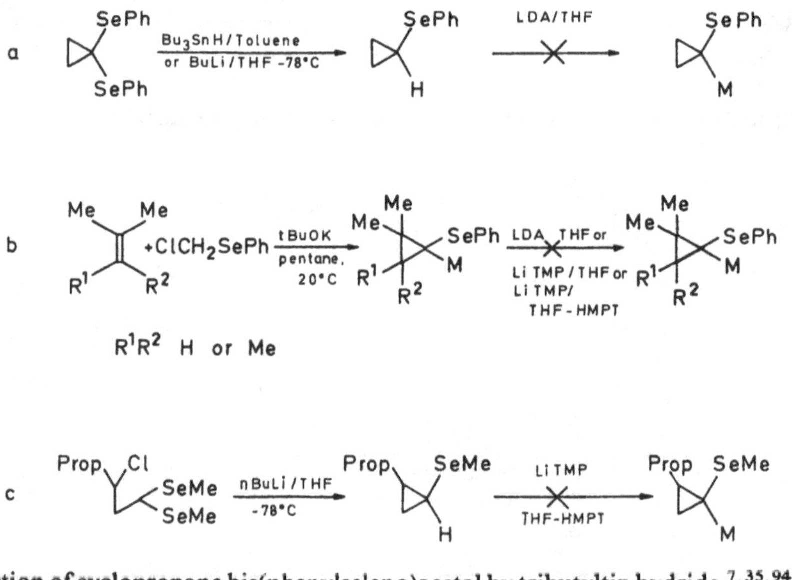

Scheme 12

a) The reduction of cyclopropane bis(phenylseleno)acetal by tributyltin hydride [7, 35, 94]) or by *n*-butyllithium [39, 94]) (*n*-BuLi) followed by protonation of the resulting α-lithio cyclopropyl selenide [7, 39]) (Scheme 12a);
b) addition of phenylselenomethylene, generated from α-chloromethyl phenylselenide and tert BuOK to alkyl substituted olefins [35, 94, 95]) (Scheme 12b),
c) The cyclisation of γ-chloro-α-lithio bisselenoacetals [35]) (Scheme 12c).

All attempts to metallate cyclopropyl silanes with strong bases such as alkyllithiums in THF [94]) or sec BuLi and TMEDA in THF [82, 98]) as well as cyclopropyl selenides with non-nucleophilic bases such as LDA in THF [39, 94]), or lithium tetramethylpiperidide in THF [35, 94]) or in THF-HMPT [35]) (Scheme 12), meet with failure.

On the other hand as expected, butyllithiums do not metallate cyclopropyl phenyl selenides. They instead act on the selenium atom producing[35] butyl cyclopropylselenides and phenyllithium rather than cyclopropyllithiums and butyl phenylselenides (Scheme 13). The metallation of cyclopropyl selenoxides has not been

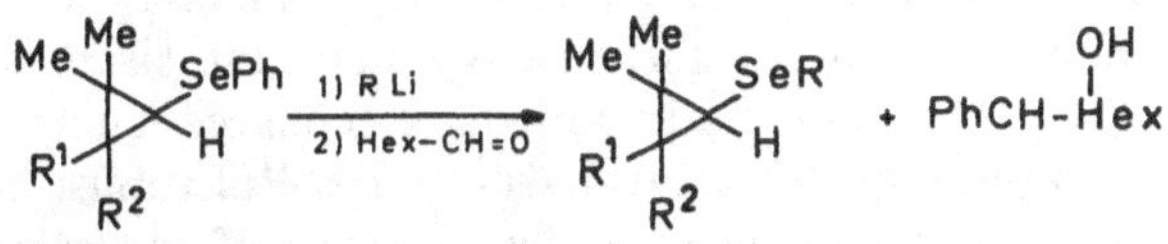

R^1,R^2 = H or Me

R = n Bu or tBu

Scheme 13

reported, but that of the cyclopropyl phenyl selenones available[67] by oxidation of cyclopropylselenides proved particularly easy and was performed[67], inter alia, by potassium *tert*-butoxide in DMSO (Scheme 14). The resulting anion, which is quite unstable, was immediately trapped[67] by benzaldehyde present in the reaction medium.

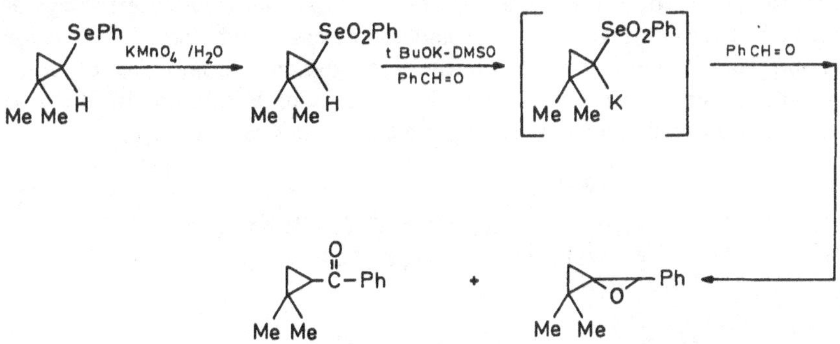

Scheme 14

2.1.2 Syntheses Implying Heteroatom-Metal Exchange

2.1.2.1 Synthesis of 1-Seleno cyclopropyllithiums by Selenium-Metal Exchange from Selenoacetals of Cyclopropanones

α-Metallocyclopropylselenides unavailable by metallation of the corresponding selenides are, however, readily available on reaction of cyclopropanone selenoacetals with alkyllithiums[35,86,87]. Although most of the work has been performed on methyl and phenylselenoacetals of the parent compound, the selenium-metal

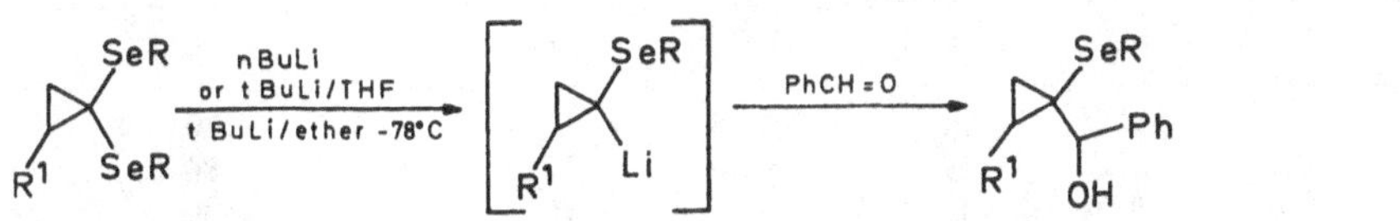

R	R^1		
R = Ph	R^1 = H		80 % (ether or THF)
R = Ph	R^1 = Pr		70 % (THF)
R = Me	R^1 = H		78 % (THF)
R = Me	R^1 = Me, hex, dec	100 %	—

Scheme 15

exchange has also been quantitatively observed with ring alkylated derivatives [35] Scheme 15.

The reaction occurs quite instantaneously at —78 °C with *n*-BuLi [86] or *tert*-BuLi [35,99] in THF, or with *tert*-BuLi in ether [87]. The availability of selenocyclopropyllithiums in the last solvent is particularly important since, for example, their nucleophilicity towards carbonyl compounds is enhanced under these conditions [87] (vide infra). However in this solvent *sec*- or *tert*-BuLi must be used [87] in place of *n*-BuLi in order to obtain quantitative cleavage of the carbon-selenium bond. For example, 1,1-bis(methylseleno)cyclopropane is recovered unchanged after addition of *n*-BuLi in ether at —78 °C or —40 °C. However, 1,1-bis(phenylseleno)-cyclopropane is more reactive since, under these conditions, 35% of 1-lithio-1-phenylseleno cyclopropane is produced [99].

It is interesting to note that the latter result is exceptional since 1,1-bis(phenylseleno)cyclopropane is the only selenoacetal derived from ketones to be at least partially cleaved under these conditions [99] and even the homologous cyclobutyl derivative is inert under these conditions. This may be due to the extra stabilization introduced by the cyclopropyl ring. The case of 2-decyl-1,1-bis(methylseleno)cyclopropane merits further comment. It is difficult to assess [35] whether the cleavage of the carbon-selenium bond occurs on the methylseleno moiety *cis* or *trans* to the alkyl group, since this organometallic leads [35] to a mixture of the two possible stereoisomers on further reaction with electrophiles (Scheme 16).

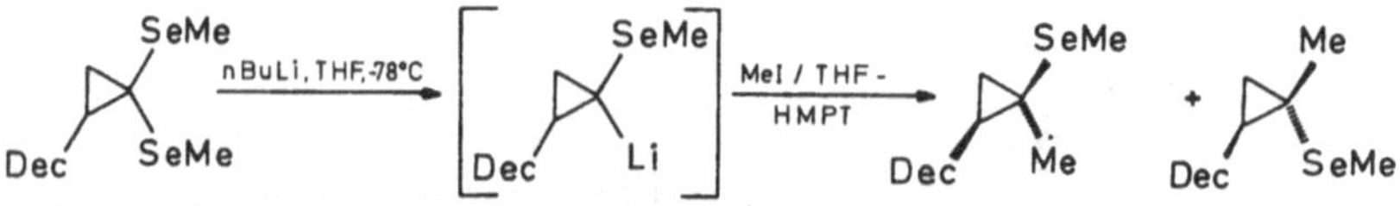

Scheme 16

2.1.2.2 Synthesis of 1-Vinyl Cyclopropyllithiums by Selenium-Metal Exchange from 1-Seleno-1-vinyl cyclopropanes

The cleavage of the carbon-selenium bond has also been used [36] for the synthesis of 1-lithio-1-vinyl cyclopropanes from 1-seleno-1-vinyl cyclopropanes (Scheme 17).

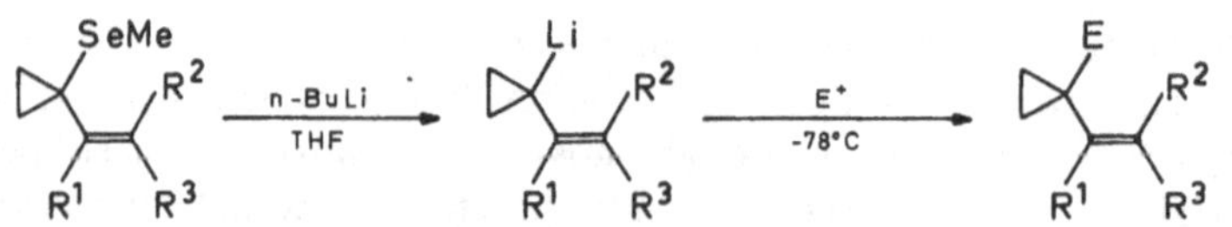

entry	R_1	R_2	R_3	Z/E	electrophile	yield
	H	H	hex	02/38	H_2O	80
	H	hex	H	90/10	H_2O	84
	hex	Peut	H	—	H	80
	H	H	hex	2/98	Dec Br	75
	H	hex	H	90/10	Dec Br	94
	H	H	hex	2/98	CO_2	60
	H	hex	H	90/10	CO_2	65

Scheme 17

These organometallics cannot, in fact, be prepared [100–102] by metallation of the corresponding carbon acid [100][1,2]. The methylseleno vinyl cyclopropanes are rapidly and regioselectively cleaved [36] by *n*-BuLi in THF from −78 °C to −45 °C, depending upon the nature of the substituents present on the carbon-carbon double bond. As far as we know, the lithium sits on the cyclopropyl carbon rather than on the other site of the allylic system, since, after reaction with water, alkylhalides, and carbon dioxide, the resulting derivatives (with the exclusion of the styryl compound) retain both the regio- and the stereochemistry originally present on the starting selenides (Scheme 17). Under similar conditions, only the *E* stereoisomer [36] is formed, whichever of the *Z* or *E* styryl compounds is reacted (Scheme 18). Phenylseleno derivatives behave differently [36] since both types of

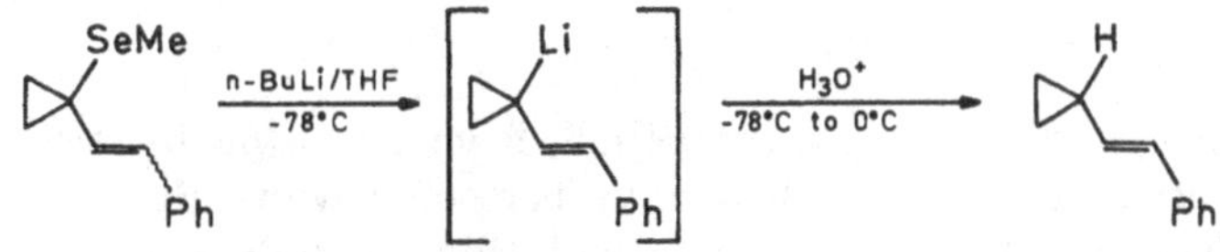

Z/E ratio 5/95 E: 76 %

Z/E ratio 98/2 E: 80 %

Scheme 18

cleavage of the carbon-selenium bond are observed, which leads [36] to a mixture, of, respectively, phenyllithium and 1-butylseleno-1-vinyl cyclopropanes, and 1-lithio-1-vinyl cyclopropanes and butyl phenylselenide. The desired 1-lithio-1-vinyl cyclopropanes can however, be exclusively formed if two equivalents of *n*-BuLi are used, [36] rather than one (Scheme 19).

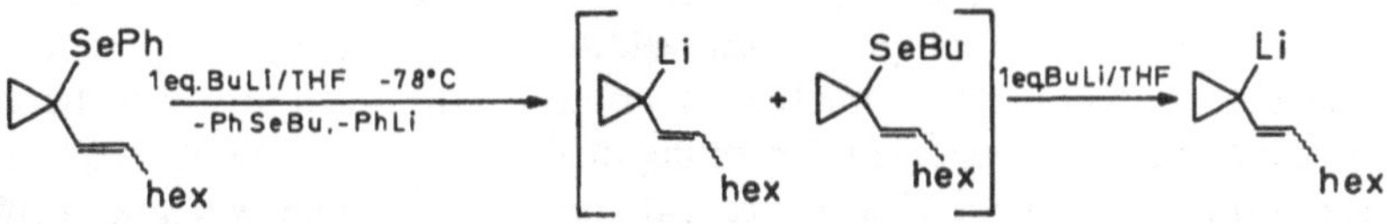

Scheme 19

2.1.2.3 Synthesis of 1-Silyl cyclopropyllithiums

By Selenium-Metal Exchange from 1-Seleno-1-silyl cyclopropanes

The selenium-metal exchange proved a valuable [77], reaction for the synthesis of α-lithio cyclopropylsilane from α-methylseleno-α-silylcyclopropane and *n*-BuLi (Scheme 20). This organometallic is in fact the first α-lithiated silane bearing two alkylsubstituents to be prepared [77,105]. The starting α-silyl selenide is readily avail-

---

¹ cyclopropyl benzene has, however, successfully been metallated[103].
They can, however be, prepared by halogen-metal exchange on 1-halo-1-vinylcyclopropanes[100, 104]

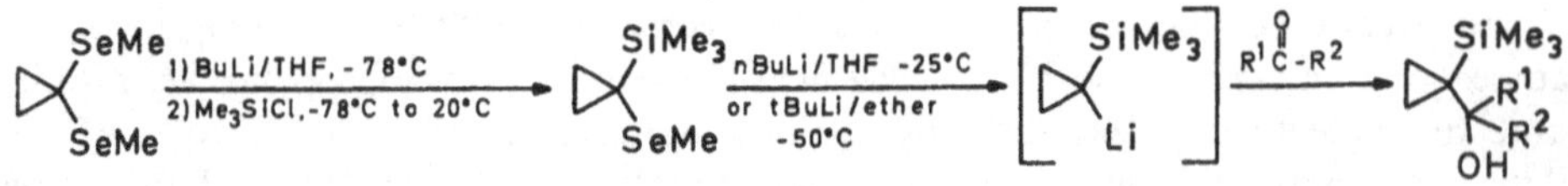

R_1	R_2	yield
H	Ph	75
H	Pr	85
H	CH(SeMe)-hex	60
H	CH=CHPr	72
Me	Non	40
H	NMe_2	80

Scheme 20

able, in the case of the parent compound, from 1,1-bis(methylseleno)cyclopropane by a sequence of reactions which involves its reaction with *n*-BuLi in THF at —78 °C and the silylation of the resulting anion with chlorotrimethylsilane. It is interesting to point out the different reactivity of α-silyl and α-selenocyclopropylselenides towards alkyllithiums: the cleavage is slow and takes place at —45 °C with the silyl derivative, whereas it occurs immediately at —78 °C with the selenoacetal. This might reflect the different stabilization of the carbanionic centers by these two different moieties.

By Sulfur-Metal Exchange from 1-Silyl-1-thio-phenyl-cyclopropanes

α-Silylcyclopropyllithium has been alternatively prepared by sulfur-metal exchange from α-thiophenyl-α-silylcyclopropane and lithium naphthalenide [82] (LN) in THF at —78 °C or with lithium 1-(N,N-dimethylamino)naphthalenide [81] (LDMAN) in THF at —50 °C (Scheme 21A). The latter conditions should be the preferred ones since dimethylamino naphthalene can be recovered easily from the crude mixture after further reaction simply by the addition of an acid to the medium. However, at least once [82] an incomplete reduction of the carbon-sulfur bond using this specific reagent was reported. Trapping of the anion with aldehydes leads [81a], in the case of the norcarane derivative, to only one stereoisomer, whereas a mixture of the two stereoisomers is formed with the lower homologs [81]. The required α-thiophenyl α-trimethyl-silyl-cyclopropanes have been prepared in two different ways (Scheme 21B) which involve either a) the silylation with chlorotrimethylsilane of 1-lithio-1-phenylthiocyclopropane prepared by metallation of phenylthiocyclopropane with *n*-BuLi [82], or by reductive cleavage of cyclopropanone bis(phenylthio)acetal with LDMAN [81], or b) the sequential treatment of 1,3-di(phenylthio)propane with two equivalents of *n*-BuLi and chlorotrimethylsilane [81, 82].

By bromine-Metal Exchange from 1-Bromo-1-silylcyclopropanes

(α-Lithio cyclopropyl)silanes bearing alkyl substituents on the ring have been conveniently prepared [77, 78] by halogen-metal exchange from (α-bromo cyclopropyl)-silanes and alkyllithiums (Scheme 22). The interest in this method lies in the accessibility of the starting material which is prepared from geminal dibromocyclopropanes

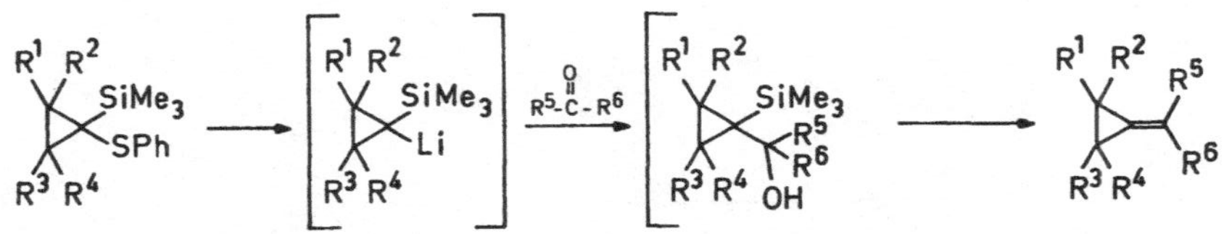

	method	R_4	R_5	yield	method	yield	Ref
	LDAMN	H	pMeOPh	85	KH/THF, 90°, 5 h	90	81 a)
	LDAMN	H	hex	84	—	—	81 a)
	LDAMN	R_4: $CH_2-CH_2-CH(tbu)-CH_2-CH_2$	:R_5	86	KH/diglyme 90°, 5 h	86	81 a)
	LN	H	c-hexyl	86	—	—	181)
	LN	$CH_2-CH_2-CH_2-CH_2-CH_2$		no yield reported			82)
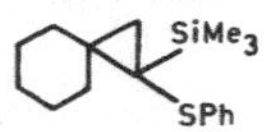	LDAMN	H	pent	92	KH/diglyme 90°, 5 h	98	81 a)
	LDAMN	H	$-C(Me)=CH_2$	90	KH/THF 25°, 1,5 h	100	81 a)
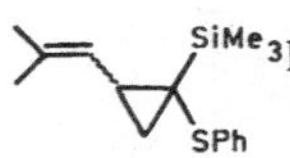	LDAMN	H	pMeOPh	92	KH/THF, 25°	95	81 a)

Scheme 21 A

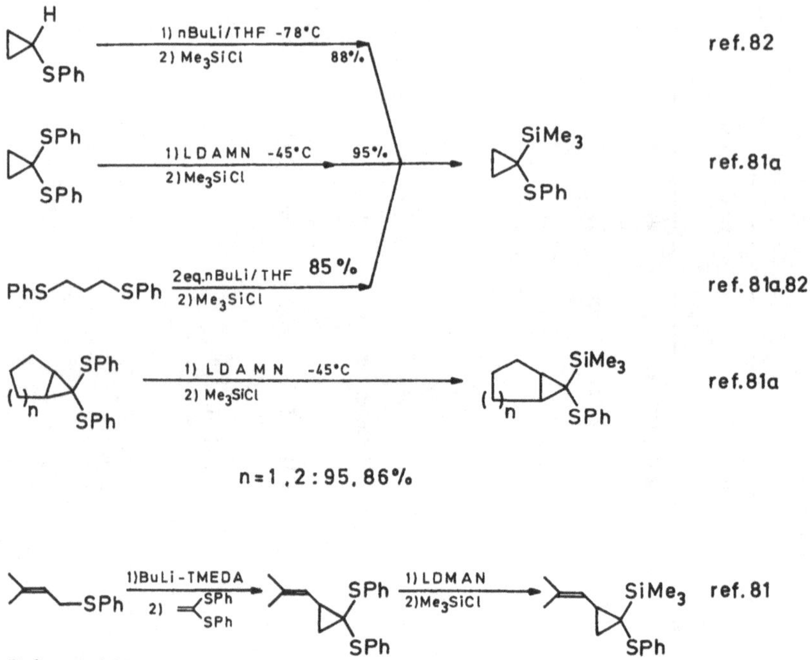

Scheme 21 B

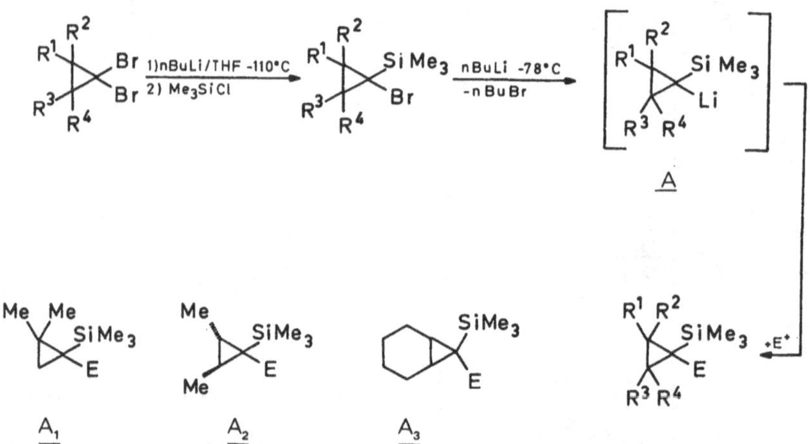

$\underline{A_1}$ $E^+ = R_5CH{=}O$ { HO–C–Dee 71% ; HO–C peut 66% ; HO–CH–CH=CH–Pr 72%

$\underline{A_2}$ HO–CH(SeMe)Ph 59% ; HO–CH–CH=CHPr 59%

$\underline{A_3}$ { HO—CH pent 40% ; HOCHCH=CHPr 10%

A E^+ = DMF {CH=O 60% refers to A_1 ; CH=O 60% $\underline{A_2}$; CH=O 10% $\underline{A_3}$

Scheme 22

by Br/Li exchange. The resulting α-bromo-α-lithio cyclopropanes have been further alkylated with chlorotrimethylsilane (Scheme 23). The halogen-metal exchange on α-

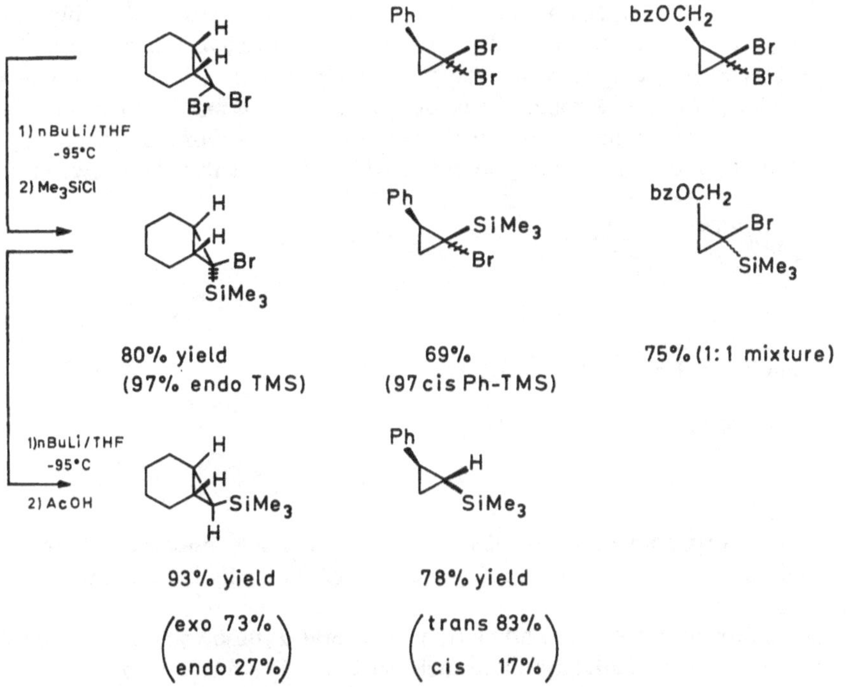

Scheme 23

bromo-α-silylcyclopropanes takes place [77,78] even at —95 °C. Even at that temperature the anions are unstable, they do not retain their stereochemistry [78] and lead to the thermodynamically more stable derivatives (Scheme 23). It is interesting to note that butyl bromide is concomitantly formed but does not interfere with the organometallic formed, which in fact is totally inert towards alkylhalides [77,78].

2.1.3 Synthesis Involving Metal-Metal Exchange

There is very little information concerning α-metallo-α-seleno or α-silyl derivatives with metals different from lithium. In two cases, however, an exchange of ligand leading to a new species containing a copper counter ion has been reported. These organocopper reagents have been used mainly to promote the allylation [35,106,107] or the acylation [78] of the cyclopropyl carbanions (scheme 24).

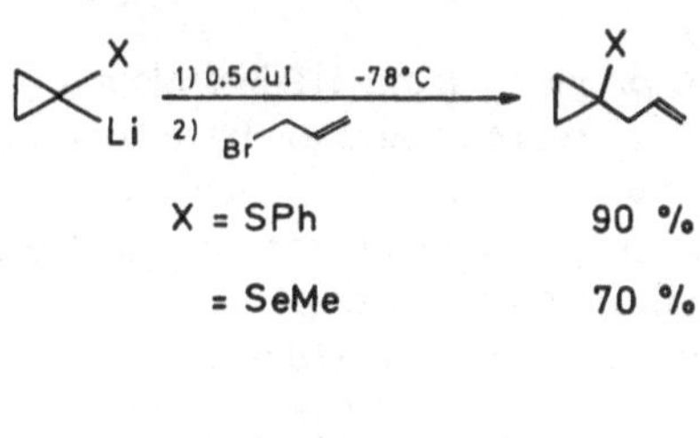

Scheme 24

For example, a new peak is observed by ^{77}Se NMR (besides the ones of α-selenocyclopropyllithium [106] after addition, at —110 °C, of small amounts (~10%) of copper (I) iodide (CuI) or, better, CuI : SMe_2-complex, which allows the formation of homogeneous solutions. This signal grows against the one of the α-seleno cyclopropyllithium when larger amounts of the complex are added, and is the only one remaining after 0.5 eq. of CuI/SMe_2 has been introduced to the medium [106].

The novel species is stable even at —50 °C; a temperature at which analogous compounds lacking the cyclopropane ring decompose to olefins [108] (Scheme 25). No effort has been made until now to extend this last reaction to cyclopropyl derivatives. Similar result have been observed with 1-lithio-1-thiophenylcyclopro-

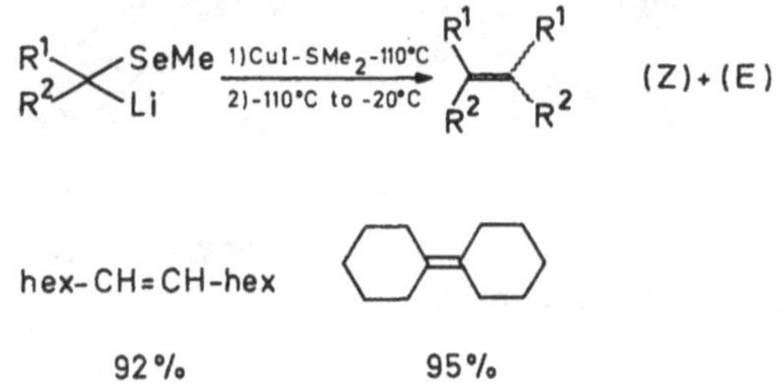

Scheme 25

pane [106, 107]. Both organometallics are allylated [106, 107] in much better yields than the lithio derivatives (Scheme 24) and react much faster with allylhalides than with carbonyl compounds.

Although, as already mentioned, alkylation of several α-lithio cyclopropylsilanes failed [77, 78], acetylation and allylation have been successfully effected [78] once lithium dibutyl-cuprate (4 eq.) has been added to the THF solution kept at —48 °C (Scheme 26).

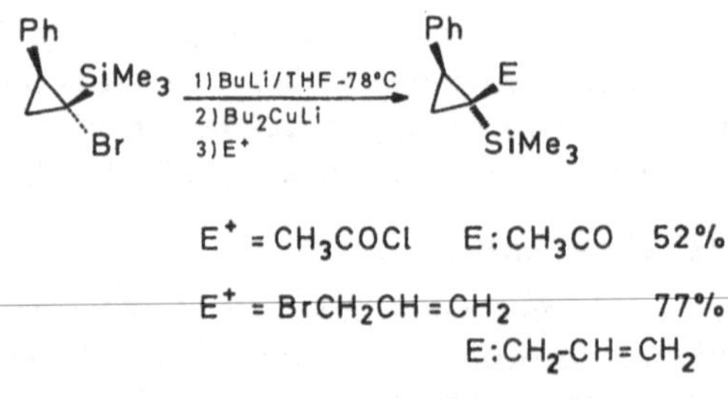

Scheme 26

2.2 Synthesis of Functionalized (1-Seleno) Cyclobutyl Metals

α-Selenocyclobutyllithiums have been prepared from 1,1-bis(seleno)cyclobutanes [57] and alkyllithiums in THF or in ether. These selenoacetals have been prepared from cyclobutanones and selenols in an acidic medium [56, 57] (Scheme 27). The method used for the synthesis of α-selenocyclobutyllithiums is identical to the one used for the preparation of other α-selenoalkyllithiums, even those bearing two alkyl groups or a cycloalkyl group on the carbanionic center [7]. These α-selenocyclobutyllithiums

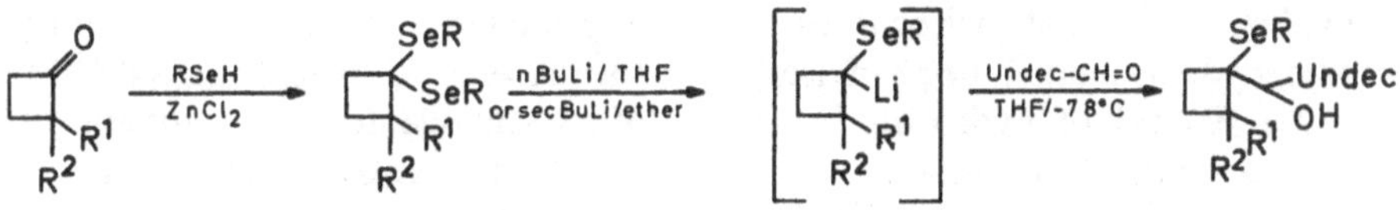

R_1	R_2	$R^*=Me$	$R^*=Ph$	
H	H	89%	84%	90% (R=Me), 71% (Ph)
H	Dec	87%	72%	—
Me	non	73%	69%	—
H	pent	74%	—	71% (R=Me)

yield %

Scheme 27

have been reacted with various electrophiles, including chlorotrimethylsilane [31]. Unfortunately, 1-seleno-1-silylcyclobutanes are not as good precursors of 1-lithio-1-silyl cyclobutanes [99] as 1-seleno-1-silylcyclopropanes are of 1-lithio-1-silyl cyclopropanes. Upon reaction with alkyllithiums, 1-methylseleno-1-silylcyclobutane leads to a mixture of unidentified organometallics, including the desired one.

2.3 Synthesis of 1,1-Bis (Seleno) Cyclopropanes

The preparation of 1,1-bis(seleno)cyclopropanes is of primary importance, since they are the precursors of α-metallocyclopropyl selenides [86,87], α-metallocyclopropyl silanes [77], and other functionalized α-metallocyclopropanes [35]. Although several synthetic methods for selenoacetals have been described [56], they are not general and that step is often the limiting one for the whole process.

The reaction involves: a) the construction of the cyclopropane ring from a selenoacetal or a selenoorthoester bearing a leaving group in the γ position, or b) the selenoacetalisation of a preexisting cyclopropane possessing the right oxidation level at one carbon, such as 1-ethoxy-1-silyloxy cyclopropane or 1,1-dihalocyclopropane.

2.3.1. Syntheses Which Involve the Construction of the Cyclopropane Ring

2.3.1.1 By Metallation Reaction

The γ-halogenoselenoacetals required for the cyclisation in the first approach have been readily prepared [86] from α, β-unsaturated aldehydes through a sequence

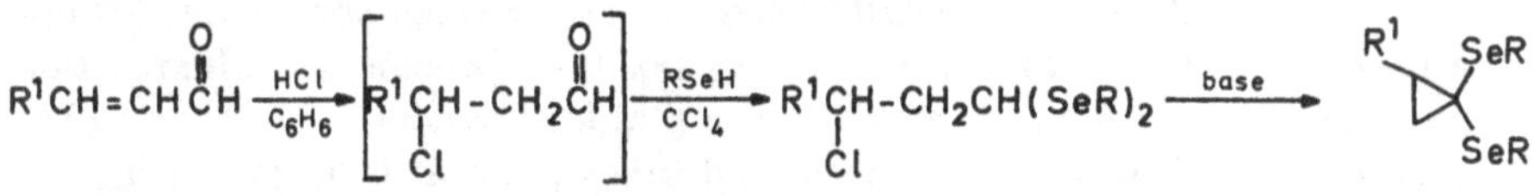

a	$R_1=H$	R=Me	MeSeH/HCl	80%	LDA, THF 0 °C	70%
b	H	Ph	$PhSeH/ZnCl_2$	75%	LDA, THF, 0 °C	80%
c	H		$PhSeH/ZnCl_2$	75%	75 tBuOK, DMSO, 20 °C	90%
d	Prop	Me	$MeSeH/ZnCl_2$	93%	LDA, THF, 0 or 20 °C	00%
e	Prop	Ph	$PhSeH/ZnCl_2$	75%	LDA, THF, 0 °C	70%

Scheme 28

which involves addition of hydrochloric acid and selenoacetalisation of the resulting β-chloroaldehyde by methyl- or phenylselenol in the presence of an acid catalyst [35,86] (Scheme 28). Excess of hydrochloric acid is often suitable for the synthesis of methylselenoacetals, but its presence must be avoided for the synthesis of their phenylseleno analogs [35,86] (Scheme 28). This is a general feature, which has already been disclosed and discussed [56] for the synthesis of other phenylselenoacetals derived from aldehydes. Cyclopropanation of the parent compounds has been routinely achieved [35,86] with 2 equivalents of LDA in THF (Scheme 28). The reaction also permits [35] the synthesis of a monoalkyl substituted phenylselenoacetal of cyclopropanone, but does not take place with the analogous monoalkylated methylselenoacetals, even when performed under more drastic conditions [35] (LDA/THF-HMPT, LiTMP/THF, or LiTMP-HMPT). This difference in reactivity between γ-halogeno substituted phenyl- and methylselenoacetals towards bases has also been observed when they are reacted with potassium *tert*-butoxide in DMSO. Under these conditions the former produce the desired cyclopropane derivative [35] (Scheme 28c), whereas the latter lead to a ketene selenoacetal [35] (Scheme 29). Presumably this reflects the different

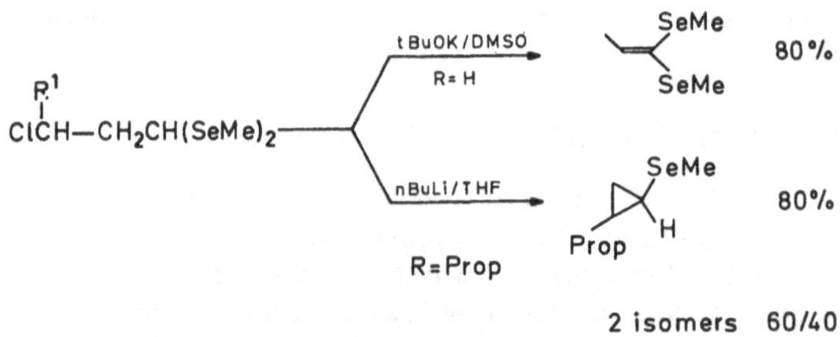

Scheme 29

acidity of the hydrogen linked to the carbon bearing the selenoacetal moiety in the two different series of compounds. The action of *n*-BuLi on γ-chloroacetals produces, as expected, the selenocyclopropane via the intermediate formation of α-lithio-γ-chloro-alkylselenide (Scheme 29b). 1,1-Bis(Phenylseleno)cyclopropane has also been prepared by Reich [54] on metallation with LDA of the monoselenoxide prepared from 3-mesyloxy-1,1-bis(phenylseleno)propane (Scheme 30 h).

2.3.1.2 By Selenium-Metal Exchange

The second approach, which involves functionalized selenoorthoesters, is reminiscent of the ones aready reported for the synthesis of thioacetals of cyclopropanone [109–111]. It permits [35,86] the synthesis of α-metallo selenoacetals bearing good leaving groups such as a tosyloxy or a mesyloxy group in the γ position (Scheme 30). The reaction takes advantage of the high nucleophilicity of α-lithioorthoesters [46–48,86] or alpha-lithioselenoacetals [46–48] towards epoxides, and takes place at 0 °C producing the γ-alkoxy selenoorthoesters [86] (Scheme 30a–g) or γ-alkoxy selenoacetals [54] (Scheme 30 h) by opening of the epoxide ring on the less substituted carbon atom. The results are obtained [86] with ethylene oxide or with terminal epoxides which react around 0 °C; the reaction with α,β-disubstituted epoxides is more sluggish (Scheme 30i). The synthesis of γ-tosyloxy selenoorthoesters has been achieved [86] by reaction of the alkoxide or the corresponding alcohol with tosyl or mesyl chloride

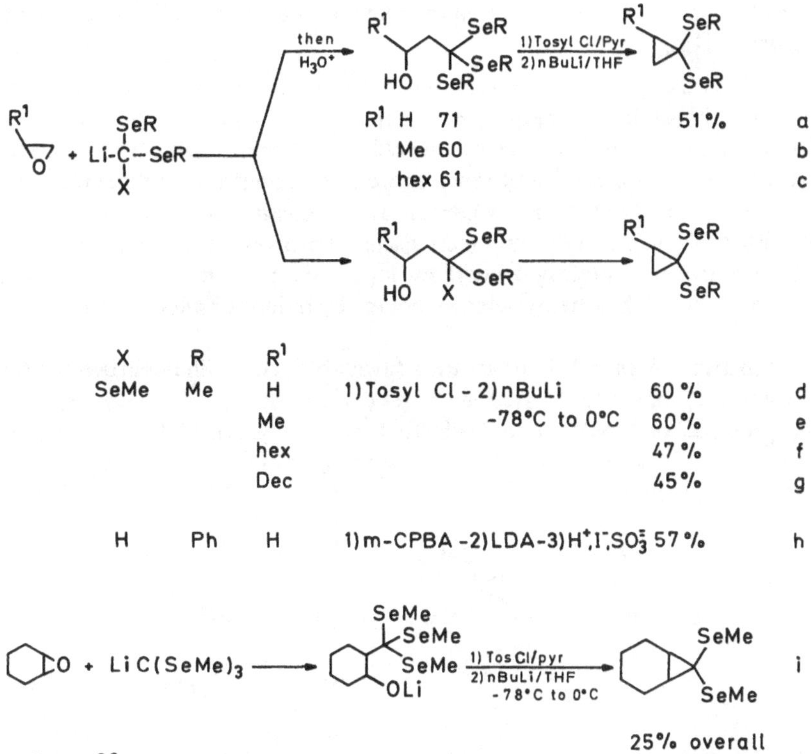

Scheme 30

(Scheme 30), and cyclisation to the selenoacetals of cyclopropanones was achieved by treatment with *n*-BuLi in THF (Scheme 30). The carbon-selenium bond of the functionalized orthoester is specifically cleaved at −78 °C but cyclisation of the resulting α-lithio-γ-tosyloxy selenoacetal takes place at a higher temperature (between −50 °C and 0 °C).

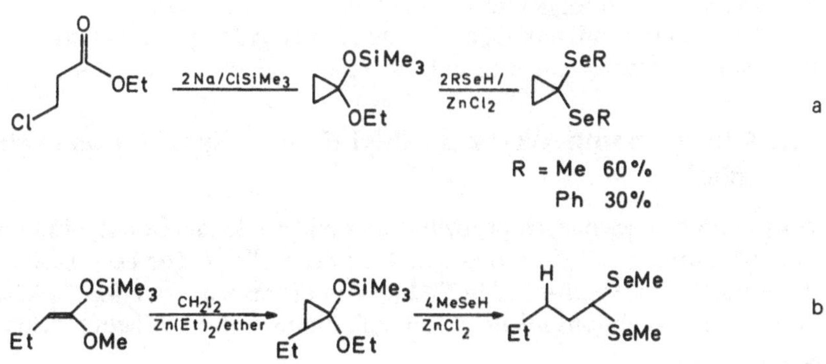

Scheme 31

2.3.2 Syntheses Which Involve the Reaction of Selenols on a Pre-built Functionalized Cyclopropane Ring

Another approach to the synthesis of the parent compounds involves [35] the reaction of 1-ethoxy-1-trimethylsilyloxy cyclopropane prepared according to the Rühlman method [112] and selenols, in the presence of zinc chloride (Scheme 31 a). The reaction is quite rapid with methylselenol [35] but much more difficult with phenylselenol [3, 35]. It takes a completely different course when an alkyl substituent is present on the cyclopropane ring [114]. Surprisingly thus methylselenol cleaves [114] the cyclopropane ring of 1-ethoxy-1-trimethylsilyloxy-2-ethyl cyclopropane [115] on addition of zinc chloride and lead [114] to 1,1-bis(methylseleno)pentane by reductive selenoacetalisation (Scheme 31 b).

Finally the substitution of 1,1dihaloalkanes (available from dihalocarbenes and olefins) by selenolates has been tested with moderate success [116]. The reaction occurs in the presence of HMPT and leads to a modest yield of 1,1-bis(methylseleno)cyclopropane on the condition that the starting material is not too highly sterically crowded [116] (Scheme 32).

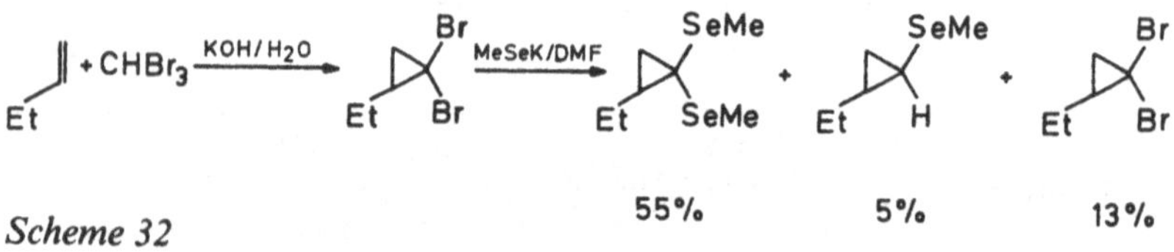

Scheme 32

3 Reactivity of 1-Functionalized-1-Metallo Small Ring Compounds

The reactivity of α-selenocyclopropyllithiums has been studied and on several occasions compared to that of their analogous α-thiocyclopropyllithiums or α-selenocyclobutyllithiums.

α-Seleno- and α-thiocyclopropyllithiums as well as α-selenocyclobutyllithiums react cleanly with primary alkyl halides [23, 35, 57] (Schemes 16, 33), allyl halides [35, 106, 107] (Scheme 24), trimethylsilyl chloride [35 77] (Scheme 33), epoxides [35, 57] (Scheme 33), and with various carbonyl compounds such as aldehydes and ketones [57, 75, 86, 87, 99], including α,β-unsaturated [35] or α-selenylated ones [31], as well as with dimethyl formamide [31, 117]. α-Silylcyclopropyllithiums, which are less nucleophilic, do not react with alkyl halides and, although they are cleanly hydroxyalkalkylated with aldehydes, they enolise ketones and lead only to modest yields of alcohols [77, 78, 81, 82].

3.1 Alkylation with Alkyl and Alklyl Halides, Epoxides, and Trimethyl silyl chloride

Alkylation of α-selenocyclopropyllithiums and α-selenocyclobutyllithiums is efficient only with primary alkyl bromides and iodides [23, 35, 57]. The best results are obtained if the reactions are performed in THF, but in the presence of HMPT (Schemes 16, 33). Under these conditions selenocyclobutyllithiums [57] lead to better yields of alkylated

---

3 for a similar result with phenylthio derivatives see reference [113]

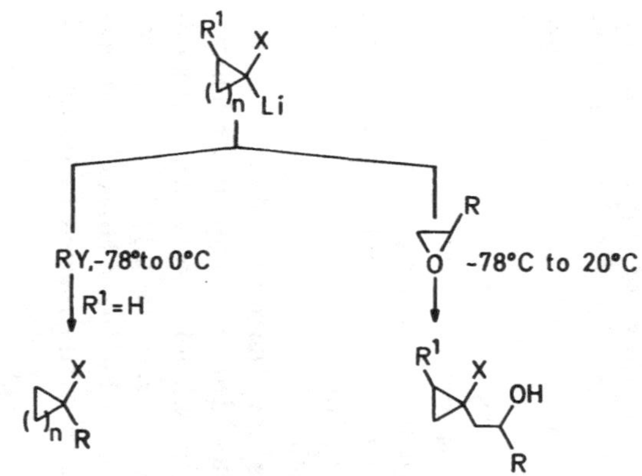

n	X	Y	R	conditions	yields	Ref.
1	MeSe	Br	Dec	THF	45	[23]
				THF-HMPT	75	[23]
1	MeSe	I	Non	THF	62	[23]
				THF-HMPT	74	[23]
1	PhSe	Br	Dec	THF-HMPT	40	[23]
		I	non	THF-HMPT	62	[23]
1	PhS	I	non	THF-HMPT	65	[23]
2	MeSe	I	non	THF-HMPT	78	[23]
2	PhSe	I	non	THF-HMPT	80	[57]
1	MeSe	Cl	Me_3Si	THF	60	[77]
2	MeSe	Cl	Me_3Si	THF	54	[35]

n	X	R_1	R	conditions	yield	Ref.
1	SeMe	H	Me	THF	64	[35]
1	SeMe	H	hex	THF	56	[35]
1	SeMe	H	hex	THF-HMPT	12	[35]
1	SeMe	hex	hex	THF-HMPT	40	[35]
2	SeMe	H	hex	THF	72	[35]
2	SeMe	H	hex	THF-HMPT	55	[35]

Scheme 33

products than their cyclopropyl [23] analogs and in the latter series methylseleno derivatives proved often to be more reactive than phenylseleno [23] or phenylthio-derivatives [23] (Scheme 33).

In the case of 2-decyl-1-lithio-1-(methylseleno)cyclopropane and methyl iodide, the alkylation leads to a 1:1 mixture of the stereoisomers [35] (Scheme 16), but it is not known which of the alkylation or the lithiation steps is not stereoselective as both stereoisomers of this cyclopropyllithium are formed. Trimethylsilyl chloride reacts efficiently with α-lithiocyclobutyl selenides [35] and α-lithiocyclopropyl selenides [77] and produces the corresponding α-silyl selenides (Scheme 23). Trimethylsilyl-(methylseleno)cyclopropame was found [77] to be a powerful precursor of α-silylcyclopropyl lithium which cannot be directly alkylated [78,94], as already mentioned.

Allylation of α-thio-[35], α-seleno- [35] and α-silyl- [35,77] cyclopropyllithiums was not very successful [35] but addition at —78 °C of 0.5 equivalent of copper (I) iodide-dimethylsulfide complex [35,106,107] prior to the allylhalide leads [35,106,107] to a very high yield of homoallyl cyclopropyl sulfides or selenides (Scheme 24). Similar observations have been made on cyclobutyl derivatives [35]. It is not clear at present whether a cuprate is involved in the process but we have evidence (^{77}Se-NMR) that a new species is transiently being formed, at least in the seleno series. The synthesis of homoallyl cyclopropylsilanes was also reported [78] and involves the allylation of a postulated cuprate formed by the addition of lithium dibutyl cuprate to α-lithiocyclopropylsilane (Scheme 26).

α-Selenocyclopropyl- and α-selenocyclobutyllithiums react with terminal epoxides [35,57] regioselectively at their least hindered site. The best results are obtained in THF. Use of HMPT as the cosolvent must be avoided since β-hydroxy selenides, arising from the ring opening of the epoxide by selenolate ions, are concomitantly formed [35] besides the expected γ-hydroxy selenides. These β-hydroxy selenides presumably occur by decomposition of the α-cyclopropyl selenide. However, we have never observed [35] the allene expected to be concomitantly formed by decomposition of the cyclopropylidene.

3.2 Hydroxy Alkylation with Carbonyl Compounds

α-Selenocyclopropyllithiums [35,66,86,87,99] and α-selenocyclobutyllithiums [57] proved particularly nucleophilic towards carbonyl compounds (Schemes 15, 27, 34). This aptitude is similar to that of α-phenylthiocyclopropyllithium [70] (Scheme 34) but by far superior to the one of α-trimethylsilylcyclopropyllithiums [77,78,81], especially when ketones are involved (Schemes 20, 21 A, 22,). It is interesting to note that the intermediary β-alkoxysilane is much less prone to decompose to olefins through a Peterson elimination reaction [77,78,81] than the analogs lacking the cyclopropane ring [118-121]. This is probably due to the strain expected to be introduced in the process. (2,2-Dimethyl-1-potassio-cyclopropyl)phenylselenone formed *in situ* with potassium tert. butoxide in the presence of benzaldehyde leads [67] directly to a mixture of the corresponding oxaspiropentane and cyclopropyl phenyl ketone (Scheme 14). This probably arises by internal substitution and hydride migration on the (β-alkoxycyclopropyl)phenylselenone transiently formed. This result [67], although limited at present to only one case, shows again [13,14] the high propensity of the selenonyl group to act as a particularly good leaving group; it resembles the ones already reported by Seebach [71] on the corresponding bromo derivative. Unfortunately no information is available at present on the nucleophilicity of (1-metallo-cyclopropyl)phenylselenones towards carbonyl compounds.

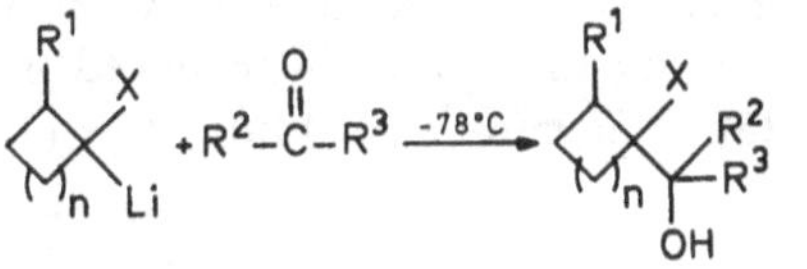

	R_1	carbonyl compound	condition	n = 0 SeMe	n = 0 SePh	n = 0 SPh	n = 1 SeMe	n = 1 SePh
a	H	benzaldehyde	THF	78	80	80	—	—
b	H	undecanal	THF	72	72	—	90	71
c	H	cyclohexyl carboxaldehyde	THF	73	63	—	—	—
d	H	undecanone-2	THF	54	—	—	78	70
e			ether	69	—	—	—	—
f	H	di-(n hexyl)ketone	THF	93	86	—	—	—
g	H	deoxybenzoin	THF	56	61	63	68	—
h			ether	70	—	—	—	—
i	H	2,2,6-trimethyl cyclohexanone	THF	47	35	45	54	—
j			ether	64	—	—	68	—
k	H	2-cyclohexanone	THF	86	—	—	—	88
l	prop	acetone	THF	—	58	—	—	—
m			ether	—	60	—	—	—

Scheme 34

α-Selenocyclopropyl- and cyclobutyllithiums are able to be hydroxyalkylated by various carbonyl compounds including highly hindered ones, such as 2,2,6,6-tetramethyl cyclohexanone, or highly enolisable ones, such as deoxybenzoin (Scheme 34). In a separate study we have tried to compare the relative nucleophilicities of α-(phenylseleno)cyclopropyllithium, its methylseleno, and its phenylthio analog towards three different carbonyl compounds, namely benzaldehyde, deoxybenzoin and trimethylcyclohexanone in the different solvents in which they can be prepared.

Competitive experiments have been made [35] in THF at —78 °C with benzaldehyde and 5 molar equivalents of phenylthiocyclopropyllithium and 5 equivalents of one of the two seleno derivatives. Recovery of equal amounts of the β-hydroxysulfide and the β-hydroxyselenide leads to the assumption that these derivatives possess similar nucleophilicities, at least towards this aldehyde.

Similar yields of β-heterosubstituted alcohols have been observed [35] with deoxybenzoin and 2,2,6-trimethylcyclohexanone for all three heterosubstituted lithio derivatives if the reactions are performed at —78 °C in THF (Scheme 34g, i), whereas the best results are obtained when (methylseleno)cyclopropyllithium is reacted at the same temperature but in ether instead of THF (Scheme 34h, j)). This is probably due to a reduction of the degree of enolisation of the starting ketone when ether is used [87]. This is a tendency which proved to be general for other α-selenoalkyllithiums [7, 8, 9, 12]

As a general trend, α-seleno- and α-silylcyclopropyllithiums have a very large tendency to add on to the carbonyl group of α-enals and α-enones. They share this tendency with other cyclopropyllithiums. This can be explained by the hardness of α-cyclopropyl metals [122]. 1-Lithio-1-seleno-cyclopropane reacts quite exclusively on the carbonyl group of 5-iodopentane-2-one, whatever the solvent used, but the nature of the product is very dependent upon the solvent: a γ-iodo alcohol is formed after hydrolysis if the reaction is performed in ether [35] whereas cyclisation of the intermediate γ-iodo alkoxide leading to a furan is produced if the reaction is performed in THF [35] (Scheme 35). 1-Seleno- [31], 1-thio- [31], and 1-silylcyclo-

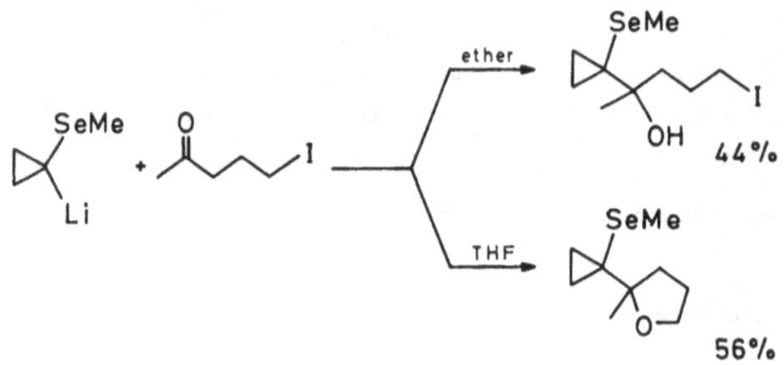

Scheme 35

propyllithiums [77] react with α-selenoaldehydes and lead predominantly to one of the possible stereoisomers of β-hydroxy selenides, whose stereochemistry can be be predicted [123] on the basis of Cram's [124] or Felkin's [125] rules. Finally α-seleno- [31] and α-silylcyclopropyllithiums [77], as well as α-selenocyclobutyllithium [35], have been

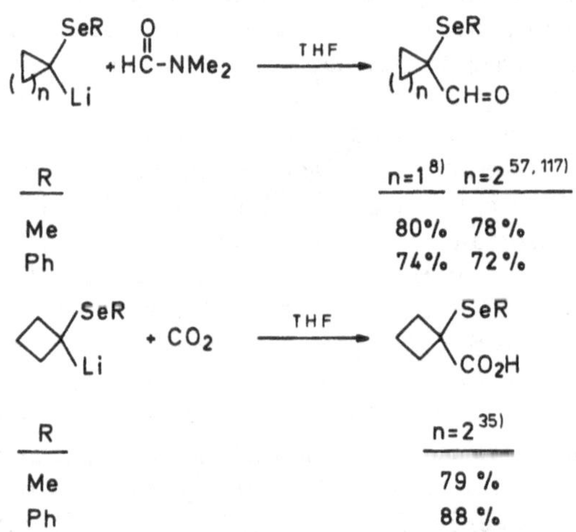

Scheme 36

successfully reacted with DMF and carbon dioxide and produce, after acidic hydrolysis, the corresponding formylcyclopropanes [31, 77] or formyl cyclobutanes [35], or the corresponding carboxylic acids [35] (Scheme 36).

4 Reactions Involving the Removal of the Selenenyl or the Silyl Moiety from α-Seleno and α-Silyl Cyclopropane and Cyclobutane Derivatives

1-Heterosubstituted cyclopropylmetals are valuable building blocks in organic synthesis, whereas 1-heterosubstituted cyclobutylmetals are practically unknown. Not only do these species introduce the heteroatomic moiety when they react with electrophiles, but they concomitantly introduce a strained cycle, which can release its strain under suitable conditions. α-Halocyclopropyl metals [72–74], cyclopropylidene (triphenyl-phosphorane) [58–60], cyclopropylidene(diphenylsulfurane) [61–62], 1-thiocyclopropyllithiums [71,79,80] and 1-isocyanatocyclopropyllithium [68,69] have been used to perform, often with greater difficulty, the transformations already possible with other members of the series missing the cyclopropane ring. This is the case for the synthesis of allylidene cyclopropanes from phosphorus ylides [58–60] or from α-halocyclopropyl metals [126,127]. This is also the case for the synthesis of oxaspiropentanes from cyclopropylidene diphenylsulfurane [61], diazocyclopropane [64] and from α-halocyclopropyl metals [71]. α-Heterocyclopropyl metals have also permitted original syntheses which utilize the strain present in the cyclopropane ring such as the synthesis of allenes [128–130] (α-halogeno derivatives) or cyclobutanones (α-halogeno [71], α-thio [11,62,131–134] derivatives). α-Heterosubstituted cyclobutylmetals were almost unknown and therefore they have not been involved in such strategies. It must however be pointed out that oxaspirohexanes usually prepared from alkylidene cyclobutanes have been used for the syntheses of a few cyclopentane derivatives (see below). It is the aim of this section to provide information relative to their selenenyl and silyl counterparts which was quite unknown six years ago:

a) to show that α-seleno- and α-silylcyclopropyllithiums, as well as α-selenocyclobutyllithiums, permit a large array of transformations, including original ones, depending upon the electrophile used and the nature of the reagent applied to the resulting compound.

b) To compare these transformation to the ones already described which use other heterosubstituted small ring compounds.

c) If unknown, to try to perform the reactions with their thio or halogen analogs.

Most of the work presented will be on the parent compounds because they are the more accessible ones. It could be extended to analogs bearing substituents on the cycloalkyl ring but this has not yet been done in all cases. Several transformations which will be reported have been previously performed on analogous compounds lacking the cyclopropane ring (see Sect. 1) but were unsuccessful when applied to the specific case of cyclopropyl derivatives. Original reagents and solutions have been found at this occasion in several instances and proved to be very useful and very efficient for those compounds lacking the cyclopropane ring.

α-Selenocyclopropyllithiums have been used, inter alia, for the synthesis of alkylidene cyclopropanes [23,45,77,87], vinyl cyclopropanes [31,36,37,77] cyclobutanones [35,87], and allylidene cyclopropanes [77,106,107] including functionalized ones. α-Selenocyclobutyllithiums have been used for the specific synthesis of cyclobutenes [57] and alkylidene cyclobutanes [57], including functionalized ones, as well as oxaspirohexanes [57,134] and cyclopentanones [134,135]. The strain present has been used for the specific synthesis of dienes [57], including functionalized ones from cyclobutenes, and for the synthesis of cyclopentanones from cyclobutanones [134,135]. α-Silylcyclopropylithiums have been used for the synthesis of alkylidene cyclopropanes [77,78,81a] allylidene cyclopropanes [77,78,81a,136], vinyl cyclopropanes [66,136–138], and cyclopentenes [137,138].

4.1 Syntheses of Alkylidene Cyclopropanes and Alkylidene Cyclobutanes

The synthesis of alkylidene cyclopropanes and cyclobutanes involving seleno compounds has been achieved via three different routes which involve
a) the formal elimination of the selenenyl moiety and a proton from 1-alkyl-1-selenocycloalkanes [23,57];
b) the formal elimination of the selenenyl moiety and hydroxyl group from β-hydroxy selenides [57,87];
c) the rearrangement of 1-selenoxy-1-vinylcyclopropanes and analogous selenonium ylides, which produces 3-hydroxy- or 3-seleno-1-alkylidene cyclopropanes [45] respectively.

Their synthesis from 1(1-silyl)cyclopropyl carbinols [77,78,81] which is closely related to the methods just presented will also be reported in (Section 4.1.2.1.2).

4.1.1 Syntheses of Alkylidene cyclopropanes and Alkylidene cyclobutanes by Formal Elimination of a Selenenyl Moiety and a Hydrogen

4.1.1.1 Syntheses of Alkylidene cyclopropanes from 1-Alkyl-1-selenocyclopropanes

Synthese of Alkylidene cyclopropanes Via the Selenoxide Route

The synthesis of olefins by oxidative elimination of selenides is one of the most versatile and useful methods in selenium chemistry [1–9,11,12]. The reaction usually takes place at room temperature with phenylseleno derivatives by simple addition of hydrogen peroxide in THF-water, sodium periodate in alcohols, or on reaction with ozone. Although the synthesis by this route of 1-cyanocyclopropene from 1(phenylseleno)-1-cyanocyclopropane has been reported [139], the cyclopropane derivatives bearing an alkyl group and a seleninyl moiety in geminal position are, as expected, more reluctant to deliver an olefin. Phenylselenoxy derivatives readily available by ozonolysis of the corresponding selenides, decompose very slowly at 110 °C in toluene to produce [23,94] (if triethyl amine as a selenic acid scavanger is present) less than 33% of alkylidene cyclopropanes after 30 hrs (Scheme 37). It is impossible to assess at present whether the hydrogen removed during the elimination reaction is the one originally present on the alkyl chain or if the reaction takes place in the ring, leading first, by removal of the more acidic hydrogen, to an alkyl cyclopropene which then rearranges to the expected more stable alkylidene cyclopropane.

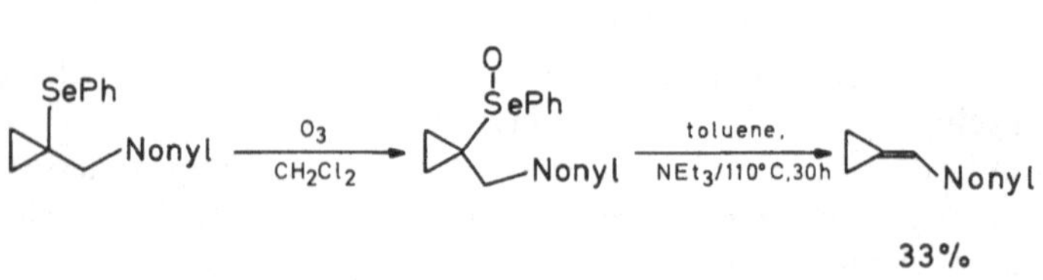

Scheme 37

Methylselenoxy analogs are even more difficult to react and treatment of the corresponding methylseleno derivative at 80 °C with *tert*-butyl hydroperoxide/basic alumina (conditions which proved particularly efficient for the synthesis of terminal olefins from methyl selenides bearing a methylseleno group at the terminus of the alkyl chain [7,8,12], which are more difficult to react) as expected [140] does not lead [35] to the desired alkylidene cyclopropanes but to low yields of cyclobutanones resulting probably from the well-known [140] reaction of *tert*-butyl hydroperoxide with the alkylidene cyclopropane formed transiently [35].

Syntheses of Alkylidene cyclopropanes Via the Selenonium route

The selenonium route proved to be more valuable. It has been specifically designed [23] by us to replace the deficient selenoxide route (Scheme 38). It was expected to produce alkylidene cyclopropanes by a mechanism which mimics the selenoxide elimination step but which involves a selenonium ylide in which a carbanion has replaced the oxide. Cyclopropyl selenides are readily transformed [23] to the corresponding selenonium salts on reaction with methyl fluorosulfonate or methyl iodide in the presence of silver tetrafluoroborate in dichloromethane at 20 °C and, as expected, methylseleno derivatives are more reactive than phenylseleno analogs. Alkylidene cyclopropanes are, in turn, smoothly prepared on reaction of the selenium salts at 20 °C with potassium *tert*-butoxide in THF [23] (Scheme 38). Mainly alkyl cyclopropenes form at the beginning of the reaction. They then slowly rearranges, in the basic medium, to the more stable alkylidene cyclopropanes [141–145] (~6 kcal/mol). In some cases the complete isomerisation requires treatment of the mixture formed in the above reaction with potassium *tert*-butoxide in THF. The reaction seems to occur via a selenonium ylide rather than via a β-elimination reaction promoted by the direct attack of the *tert*-butoxide anion on the β-hydrogen of the selenonium salt, since it has been shown in a separate experiment [23] that the reaction does not occur when a diphenylselenonium salt (unable to produce the expected intermediate) is used instead of the phenyl-methyl or dimethyl selenonium analogs. It has also been found that the elimination reaction is the slow step in the process, since styrene oxide is formed if the reaction is performed in the presence of benzaldehyde which traps the ylide intermediately formed [35].

A similar reaction takes place with phenylthio derivatives and therefore the alkylidene cyclopropanes can be prepared (Scheme 38) from cyclopropanone-selenoacetals or cyclopropyl-phenylsulfide in reactions which involve

1) reaction with *n*-BuLi which produces the α-heterosubstituted cyclopropyllithiums which are further alkylated with primary alkylhalides;

2) the alkylation of the resulting selenides or sulfides with a methylating agent and further treatment with potassium *tert*-butoxide of the selenonium/sulfonium salt.

It should be recalled that the reaction does not work with secondary alkylhalides and that the methylseleno derivatives offer the advantages over the others of the volatility of the byproduct dimethyl selenide formed concomitantly. This permits the facile purification of the olefin produced. This type of reaction has been successfully adapted, with minor changes, to the preparation [35,106,107] of allylidene cyclopropane; a valuable diene in Diels Alder reactions [107] (Scheme 38). The α-heterosubstituted cyclopropyllithiums have been allylated in high yields with allyl halides, on the addition of a copper (I) iodide dimethylsulfide complex to the

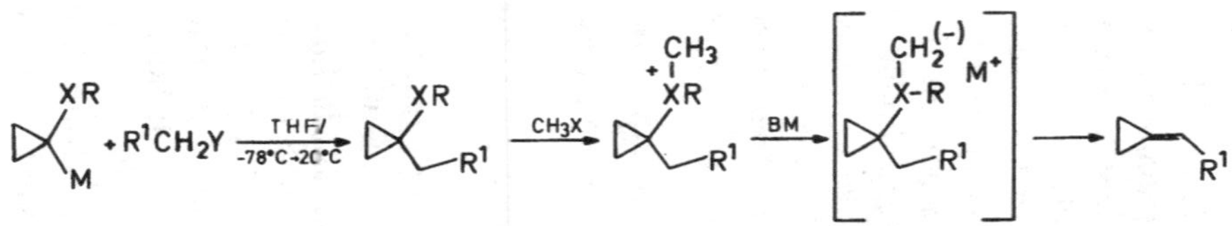

Rx	R$_1$	Y	additive						Ref.
MeSe	octyl	I	HMPT	74	$CH_3I-AgBF_4/CH_2Cl_2$	tBuOK/THF 20°, 18 h	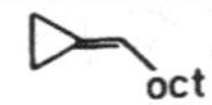	77%	91)
PhSe	octyl	I	HMPT	62		40 h		48%	91)
PhS	octyl	I	HMPT	65		40 h		42%	91)
PhS	vinyl	Br	CuI	80	$MeSO_3F$ neat	KOH/DMSO 20 h		70%	106)
MeSe	vinyl	Bu	CuI	70	$MeSO_3F$ neat 20° or Me_2SO_4 60°	tBuOK/DMSO	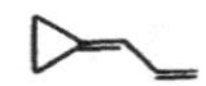	68%	85, 104)

Scheme 38

medium. The selenonium/sulfonium salts have been prepared as described above [106,107] but now the elimination of the selenide/sulfide is easier due to the presence of an extra double bond. Thus potassium hydroxide in DMSO is strong enough to provide [106,107] excellent yields of the allylidene cyclopropane (Scheme 38) the reactivity of which will be discussed in Sect. 4.4.

4.1.1.2 Syntheses of Alkylidene cyclobutanes from 1-Alkyl-1-selenocyclobutanes

Both the selenoxide and the selenonium ylide routes have been applied to cyclobutyl derivatives, themselves readily available [57] from selenoacetals of cyclobutanones on one hand and primary alkyl halides, epoxides, or carbonyl compounds on the other.

Syntheses of Alkylidene cyclobutanes Via the Selenoxide Route

The selenoxide route which was particularly inefficient with cyclopropyl derivatives in this case proved suitable. The reaction is completely regioselective in the case of β-hydroxy selenides, which produce exclusively the allyl alcohols possessing the endocyclic double bond (Scheme 39) [57], whereas a mixture of endo and exo olefins is

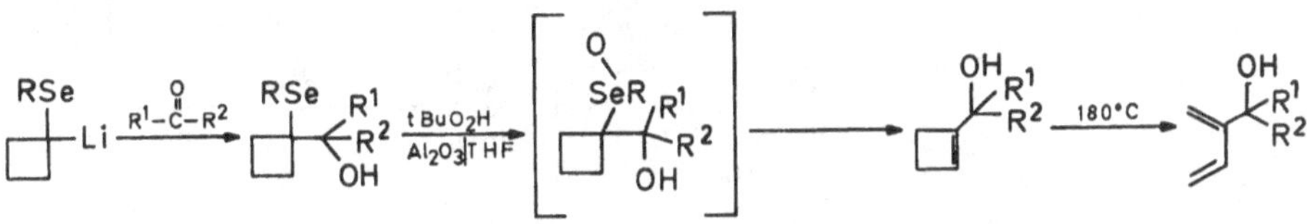

A B C D

R	R_1	R_2	A*	C*		D*
Ph	H	Decyl	71%	70%	:	75%
Me	H	Decyl	90%	70%	:	75%
Ph	Me	nonyl	70%	76%	:	85%

* refers to the yields

Scheme 39

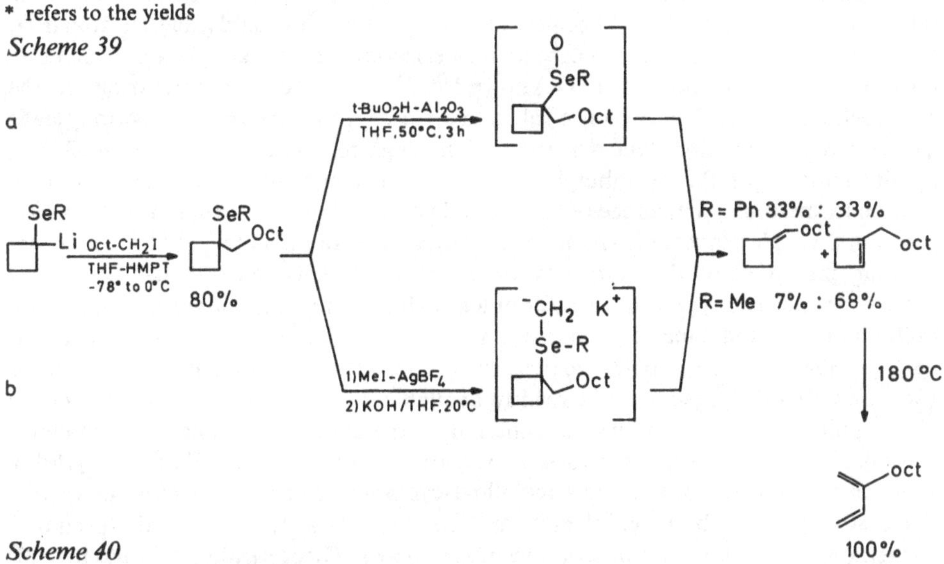

Scheme 40

formed (often with a predominance of the exocyclic one) when applied to 1-alkyl-1-selenoxycyclobutanes [57] (Scheme 40a) and to 1-(seleno)-1-(2'-hydroxyalkyl)cyclobutames [57] (Scheme 41a). Since cyclobutenes are more stable than alkylidene

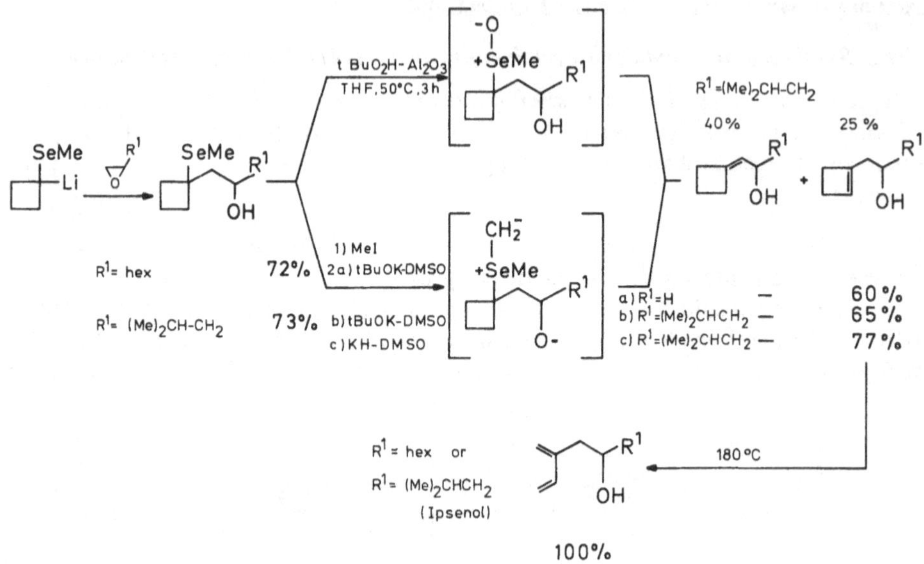

Scheme 41

cyclobutanes [146], these results suggest that these reactions are, at least partially, under kinetic control.

Syntheses of Alkylidene cyclobutanes Via the Selenonium Route

Much greater regioselectivity in favor of the endocyclic isomer is observed when the selenonium route is used [57] (Schemes 40b, 41b). This is probably due, although no experimental proof has been given, to an isomerisation which takes place concurrently during the process and which is known [146–150] to favor the formation of the endocyclic isomer. The presence of a hydroxy group in the γ position leads exclusively to the cyclobutene derivatives. This high regioselectivity can be explained by the removal of the cyclobutyl hydrogen, which occurs under kinetic control, by an intramolecular assistance of the alkoxide anion, thus promoting the elimination reaction through a favorable six-membered cyclic transition state [35,57,151], or by enhancing the speed of the isomerisation reaction [35]. This reaction has been used for the connective synthesis of cyclobutenes, including functionalized ones, from cyclobutanones and, since cyclobutenes are thermally labile [152–162], it has permitted [57] a powerful entry to the regioselective synthesis of 2-substituted butadienes (Schemes 39–41). These are not readily available from 2-metallobutadiene and an electrophile, due to difficulties encountered in the synthesis of this organometallic intermediate. In our approach, α-selenocyclobutyllithiums, available from cyclobutanones, play the role of masked 1-metallo-1-cyclobutenes or of 2-metallobutadienes. This strategy has been efficiently applied to the synthesis of dl Ipsenol [57] (Scheme 41), an aggregative pheromone of *Ips Barkae*. Theoretically, it should provide an elegant synthesis of the optically pure Ipsenol since the chiral center present on the epoxide should not be touched in the process.

4.1.2 Syntheses of Alkylidene Cyclopropanes and Cyclobutanes by formal Elimination of a Hydroxyl Group and a Heteroatomic Moiety

The methods discussed above, although efficient, possess important limitations. They do not permit, for example, the synthesis of tetrasubstituted alkylidene cyclopropanes due to the unavailability of the starting selenides, the alkylation of α-selenocyclopropyllithiums with *sec*-alkylhalides being not feasible at present.

4.1.2.1 Syntheses of Alkylidene cyclopropanes

Syntheses of Alkylidene cyclopropanes from β-Hydroxyalkyl selenides

In general, β-hydroxyalkyl selenides are powerful precursors to olefins. The reaction is usually carried out [4–9, 11, 12] by simple mixing of the selenium derivative with thionyl chloride, mesyl chloride, phosphorus oxychloride, and trifuoroacetic anhydride in the presence of triethylamine and already takes place at 20 °C. It probably involves [163] the selective transformation of the hydroxyl group of β-hydroxy selenides to a better leaving group, which is followed by the formation of a seleniranium ion. This further loses the selenenyl moiety by the attack on the selenium atom of the counter ion or of the triethyl amine acting as a nucleophile and produces the olefin. The reagents reported above have been used without problems for the synthesis of several alkylidene cycloalkanes [4–9, 11, 12], including alkylidene cyclobutanes [57], but proved quite inefficient for the preparation of the cyclopropylidene analogs [87]. It was expected that alkylidene cyclopropanes would be more difficult to prepare than higher homologs, due in particular to the strain introduced [147, 164] during the transformation and present in the olefin (Scheme 42). Side reactions may also occur, due to the high reactivity of such strained olefins towards the species present in the reaction medium and also due to the well-known propensity of cyclopropylcarbinols to rearrange to cyclobutyl [70, 165] or/and to homoallyl [166, 167] derivatives when the hydroxyl group is transformed to a better leaving group. In fact, the reagents reported above are not convenient for the synthesis of alkylidene cyclopropanes from β-hydroxyselenides. It was also found that the substitution on the carbon bearing the hydroxyl group has a great influence on the

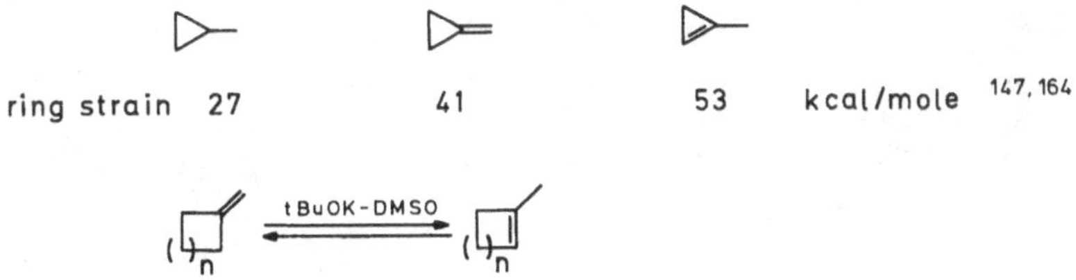

n	1	2	3	4
relative rate	1070	454	1	5.8 [146]
endo/exo+endo ratio at the equilibrium	86	99.91	99.6	98.7

Scheme 42

nature of the products formed. Those bearing a tertiary alcohol functionality lead to 1-seleno-1-vinylcyclopropanes in reasonable yield when thionyl chloride/triethylamine is used [37] and to low yields of the alkylidene cyclopropanes on reaction with phosphorus oxychloride/triethylamine [35], whereas several unidentified products are obtained from those bearing a tertiary alcohol, functionality what ever the reagent used [37].

Phosphorus triiodide or diphosphorus tetraiodide/triethylamine at 20 °C (Scheme 43) carbonyl diimidazole or thiocarbonyl diimidazole/toluene at 110 °C, (Scheme 44) however, cleanly provide [87] alkylidene cyclopropanes from methylseleno derivatives. The best results are obtained with the first reagent for those (β-hydroxy-alkyl-methylselenides derived from aldehydes (Scheme 42) whereas PI_3/NEt_3 works parti-

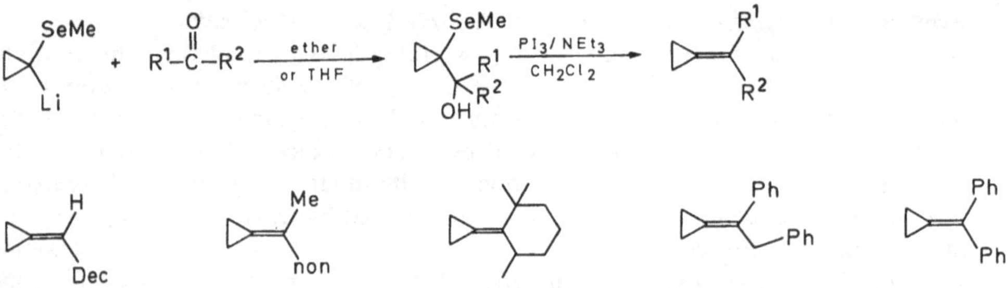

72*,30 69*,65 64*,75 70*,75 90*,74

Scheme 43

\+ The first number refers to the yield in β hydroxyselenide, the second to the yield in alkylidene cyclopropane

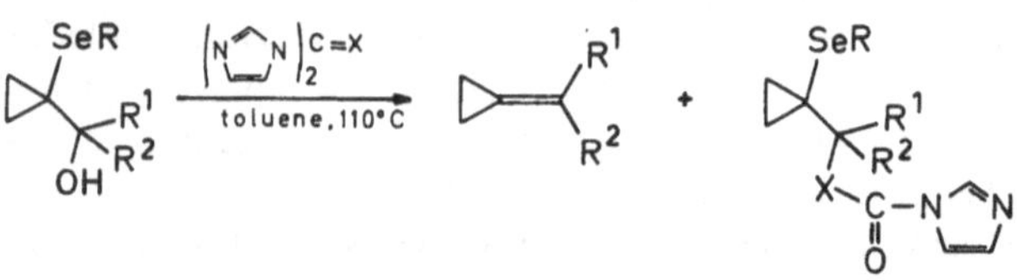

A B

	R	R_1	R_2	X	A	B
a	Me	Dec	H	S	31	51
b	Me	Dec	H	O	65	0
c	Me	Ph	H	O	70	
d	Me			O	72	
e	Me		H	O	82	
f	Ph	Dec	H	O	00	55
g	Ph	Ph	H	O	30	
h	Me	Non	Me	O	33	

Scheme 44

cularly well for those derived from ketones (Scheme 43). The nature of the substituent linked to the selenium atom also has a great influence on the success of the transformation, the methylseleno derivatives being much better precursors of alkylidene cyclopropanes than their phenylseleno analogues. The strategy described for the synthesis of alkylidene cyclopropanes from carbonyl compounds is very efficient and compares very well with other methods, such as the Wittig reaction introduced in this specific field by Bestman [59]. It permits [87], inter alias, the synthesis of alkylidene cyclopropanes derived from highly enolisable carbonyl compounds such as deoxybenzoin or from quite hindered carbonyl compounds such as 2,2,6-trimethylcyclohexanone (Scheme 43), which do not seem to be available by the Wittig reaction. The higher aptitude of 1(1-methylseleno)-1-(2′-hydroxyalkyl) cyclopropanes compared to their phenylseleno analogues to produce alkylidene cyclopropanes has been used for the selective (100%) synthesis [35] of 1-(1-phenylseleno)-1-(1′-cyclopropylidene) cyclopropanes from 1-(1-phenylseleno) cyclopropyl 1-(1-methylseleno) cyclopropyl carbinols (scheme 44d).

On the other hand, as already pointed out 1-(1-seleno)-1-(2′-hydroxyalkyl) cyclopropanes are much less pronte to produce olefins than their open chain analogues (Scheme 45). This particularity has been used, as it will be shown in Sect. 4.2.2 (Schemes 59, 60) for the synthesis [31] of 1(1-seleno)-1-vinylcyclopropanes from 1-(1-seleno)-1-1′hydroxy-2′-selenoalkyl) cyclopropanes.

This difference in reactivity is particularly significant since 1-(1-methylseleno)-1-vinylcyclopropanes are formed even [31] from 1-(1-methylseleno)-1-(1′-hydroxy-2′-(phenylseleno)alkyl)cyclopropanes (Scheme 45b, f): for the first time the phenylselenenyl rather than the methylselenyl moiety is removed in the process.

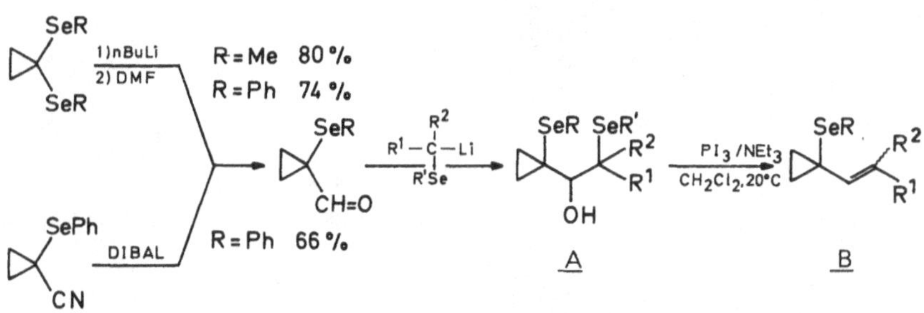

	R	R′	R_1	R_2	A	B
a	Me	Me	H	hex	96%	93% (55)*
b	Me	Ph	H	hex	63%	81% (70)
c	Ph	Me	H	hex	86%	80% (55)
d	Ph	Ph	H	hex	89%	91% (80)
e	Me	Me	H	Ph	95%	72% (68)
f	Me	Ph	H	Ph	75%	78% (75)
g	Me	Me	Me	nonyl	42%	96% (55)*
h	Me	Me	hex	hex	81%	90%
i	Ph	Ph	$-(CH_2)_3-$		72%	92%

* These numbers refer to the percentage of the major isomer the stereoisomer assignement has not yet been achieved

Scheme 45

These 1-seleno-1-vinylcyclopropanes have proved especially useful, inter alia, for the synthesis of allyl alcohols where the carbon-carbon double bond is part of an alkylidene cyclopropane [45] and for the preparation of allylidene cyclopropanes [45], (see page 52).

The novel reagents which have been introduced for the synthesis of alkylidene cyclopropanes from (β-hydroxyalkyl) methylselenides proved particularly efficient for the synthesis of a large variety of olefins, including other alkylidene cycloalkanes[57]. PI_3 or P_2I_4/NEt_3 are also to transform (β-hydroxyalkyl) sulfides to olefins[168, 169] but instead lead[170] to a stable 1-(1-phenylthio)-1-(iodomethyl) cyclopropane starting from 1-(1-phenylthio)-(hydroxymethyl)cyclopropane (scheme 46) The reaction of

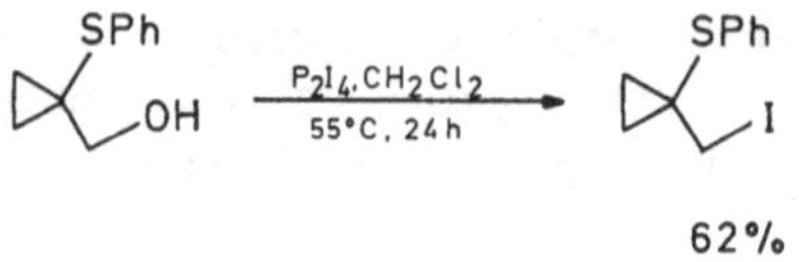

Scheme 46

PI_3/NEt_3 with 1-(1-methylseleno)-(1′-hydroxy-2′alkenyl)cyclopropanes takes a different course. The iodide ion attacks the cyclopropane ring and produces[87] the unstable diene bearing a primary alkyl iodide (scheme 47a). Similar behavior of PI_3 was also observed[77] with 1-(1-silyl)cyclopopyl carbinols (Scheme 47b)

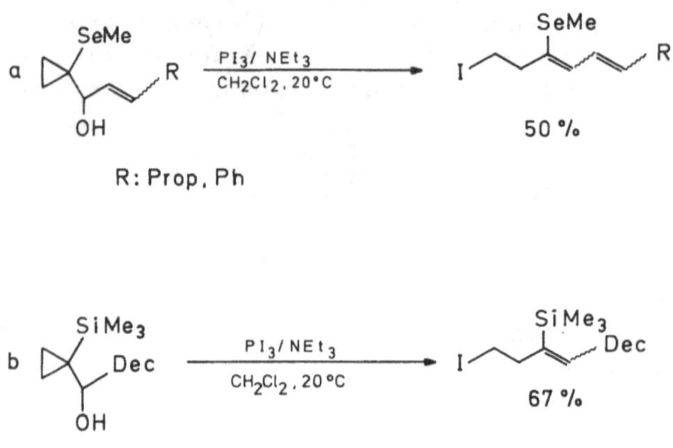

Scheme 47

Synthesis of Alkylidene and Allylidene cyclopropanes from 1-(1-silyl) cyclopropyl carbinols

The synthesis of alkylidene and allylidene cyclopropanes reported in this section takes advantage of the availability [77, 78, 81a, 82] of 1-(1-silyl) cyclopropyl carbinols from α-lithio cyclopropylsilanes and carbonyl compounds. It, however, suffers from the sometimes modest yields obtained when ketones are involved (Schemes 21a, 47) in the Peterson olefination reaction [77,78,81a] (Schemes 21, 48). This reaction seems much more difficult to achieve than when straight-chain analogs are involved and resembles the cases of allenes [121] and chlorocyclopropenes [120] reported by Chan. For example, thionyl chloride alone is not suitable for that purpose [77,136] but further addition of tetra-*n*-butylammonium fluoride (20 °C, 15 hrs) leads to the formation of undecylidene cyclopropane [77,136] in 46% yield from the corresponding 1-(1-silyl)cyclopropyl

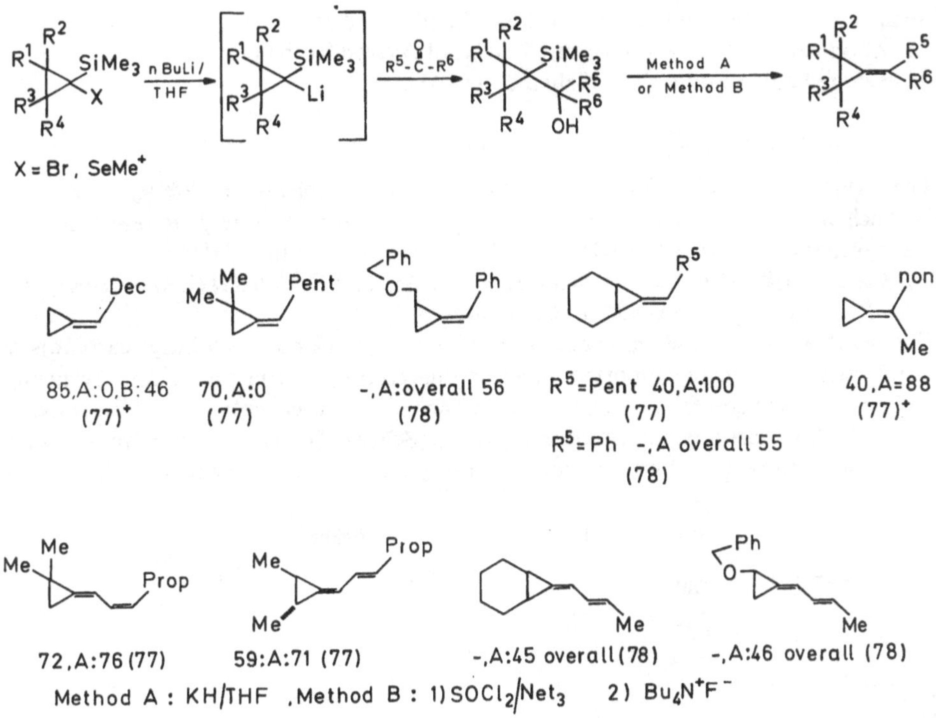

Scheme 48

informations below formula refer to the yield in β hydroxysilane, method used for the olefination reaction yield of this step or overall yield (reference)

\+ except these to olefins which have been obtained from 1-seleno-1-silyl-cyclopropane the olefins have been prepared from 1-bromo-1-silylcyclopropane.

carbinol (scheme 48) or of dicyclopropylidene ethane [136] in 76% yield (Scheme 49).

The transformation of 1-(1-silyl)cyclopropyl carbinols to alkylidene cyclopropanes is not efficiently achieved with KH in THF [77,81,136] under conditions usually used for straight chain analogues. The results are unpredictable; the yields in alkylidene cyclopropane being often very low [77] but in some cases very good too [77] (Schemes 21 A, 48). Better yields are however, observed, when the reactions are per-

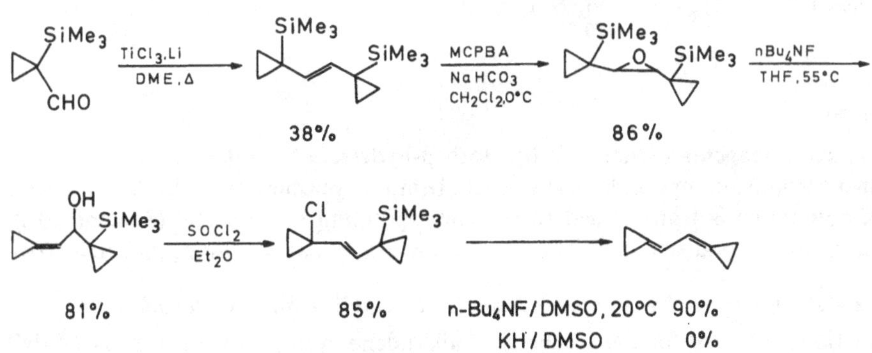

Scheme 49

formed in diglyme of 90 °C, on the other hand the reaction seems to be more facile, when applied to 1-(1-silyl)cyclopropyl carbinol derived from ketones or α,β-unsaturated carbonyl compounds [77,78] (Scheme 21a, 48).

4.1.2.2 Syntheses of Alkylidene Cyclobutanes

The synthesis of alkylidene cyclobutanes from 1-(-seleno) cyclobutyl carbinols is much more facile than of those bearing a cyclopropyl moiety. It resembles the transformation of other (β-hydroxyalkyl) selenides to olefins [7-9,11,12].

Synthesis of Alkylidene Cyclobutanes from 1-(1-Seleno) Cyclobutyl Carbinols and from 1-(1-Hydroxy)-1-(1′-selenoalkyl) Cyclobutanes

The synthesis of alkylidene cyclobutanes from 1-(1-seleno) cyclobutyl carbinols is much easier than the one reported for analogous compounds with a cyclopropyl group and readily occurs on their reaction with a large variety of reagents [7-9,11,12] especially PI_3 at 20 °C or carbonyl diimidazole at 110 °C (Scheme 50k) Alkylidene cyclobutanes have also been prepared on reaction of 1-(1-hydroxy)-1-(1′-selenoalkyl)cyclobutanes

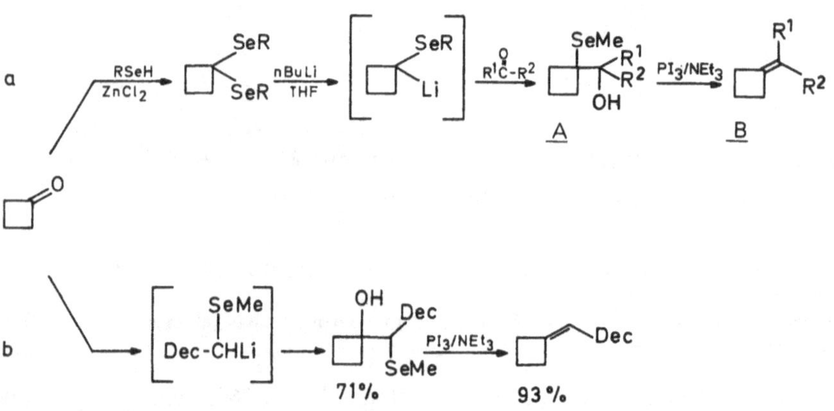

R_1*	R_2	A, yield %	condition	B, yield %
H	octyl	78%	$Im_2C{=}O$, Toluene 110 °C	91%
H	octyl	90%	PI_3/NEt_3, CH_2Cl_2 20 °C	70%
Me	nonyl	78%	PI_3/NEt_3, CH_2Cl_2 20 °C	87%

* refers to reaction a

Scheme 50

with the same reagents (Scheme 50b). Both β-hydroxyalkyl selenides are available from two carbonyl compounds, one of them being a cyclobutanone. In the first case, the cyclobutanone is transformed to the corresponding selenoacetal (Scheme 50a), whereas in the second one it is reacted with an α-selenoalkyllithium (Scheme 50b).

Synthesis of Alkylidene Cyclobutanes from 1(1-Silyl) Cyclobutyl Carbinols

The reactions reported for the synthesis of alkylidene cyclopropanes from 1-(1-Silyl) cyclopropyl) carbinols do not apply to the whole series of alkylidene cycloalkanes but seem at present to be restricted to cyclopropyl derivatives[77]. The other approach involving a (β-hydroxyalkyl) silanes/derived from the reaction of an (α-metalloalkyl) silane and a cyclobutanone does not seem to have been described.

4.2 Syntheses of Vinylcyclopropanes

The mutual interaction between the carbon-carbon double bond and the cyclopropane ring of vinylcyclopropanes confers to this class of compounds a particularly high reactivity which has been used, inter alia, for the synthesis of cyclopentenes[171–173] or 1,4-dienes [174,175]. Ring functionalized derivatives are also known for adding some organometallics [176–179] including organocuprates [177,179] or radicals [178,180]. Depending upon the reagent used and the nature of the starting material, the cyclopropane ring can remain or can be completely destroyed. Only a few derivatives bearing a heteroatomic moiety on the cyclopropane ring and at the allylic position had been prepared prior to our work. The synthesis of the bromo[127] and the phenylthio[70] derivatives must be recalled, although no special attention has been paid to the control of the stereochemistry on the carbon-carbon double bond. 1-seleno-1-vinyl cyclopropanes and *1*-silyl-1-vinyl cyclopropanes were unknown five years ago. We have set up a series of methods which have allowed the synthesis of several members of the series, including those bearing a bromine [181] atom, a sulfenyl [45], a selenyl [31,37] or a silyl [77] moiety in the 1-position. We also found that these compounds can play a crucial role in the synthesis of various cyclopropyl and cyclobutyl derivatives, inter alia vinylcyclopropanes bearing a metal at the position where the heteroatom was attached[37], functionalized alkylidene cyclopropanes[37] and cyclobutanes[37], as well as allylidene cyclopropanes [45,136]. Cohen [81a] and Paquette [82,136–138,182] have also prepared several vinylcyclopropanes bearing a silyl group at the ring junction and have used them for the synthesis of functionalized cyclopentenes [173,138] and of dicyclopropylidene ethane [136].

4.2.1 Synthesis of 1-Hetero-1-vinylcyclopropanes by Dehydration Reactions

4.2.1.1 Synthesis of 1-Seleno-1-Vinylcyclopropanes

The first approach to the seleno series taken advantage of the wide availability of 1-(1-seleno)cyclopropyl carbinols and the small tendency (discussed in Section Syntheses of alkylidene cyclopropanes from β-hydroxy alkyl selenides), of the phenylseleno derivatives to produce alkylidene cyclopropanes on reaction [37], with for example, thionyl chloride/triethylamine mixture (Scheme 51). However, the best results have

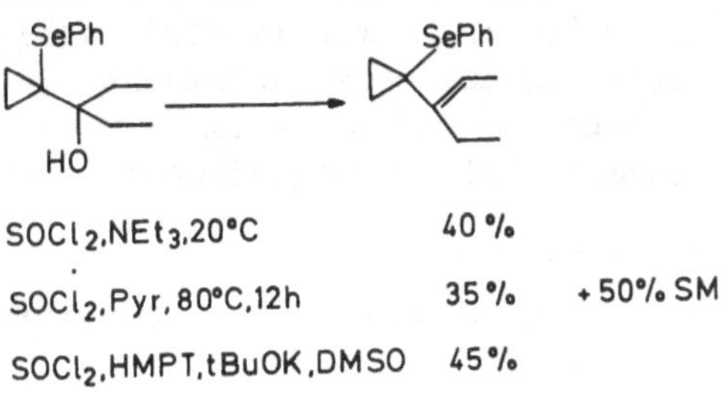

Scheme 51

been obtained with the Burgess reagent ($MeO_2CN^- SO_2 NEt_3$), [183, 37] the mixture being refluxed in toluene for 2 to 40 h (Schemes 52—54). This method, first described by Burgess [183] then used by Trost [165] for the dehydration of cyclopropyl carbinols

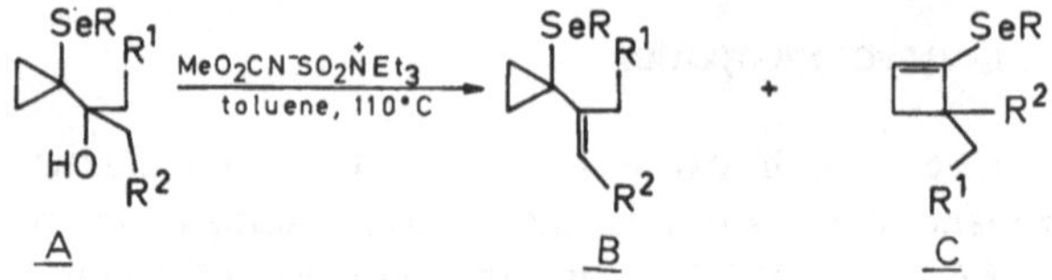

R	R_1	R_2	time required (hr)	overall yield	$\frac{B}{B+C}$ %	$\frac{C}{B+C}$ %
Ph	Et	Et	3	80%	82	18
Me	Et	Et	2,5	60%	90	10
Me	hex	hex	15	90%	95	05

Scheme 52

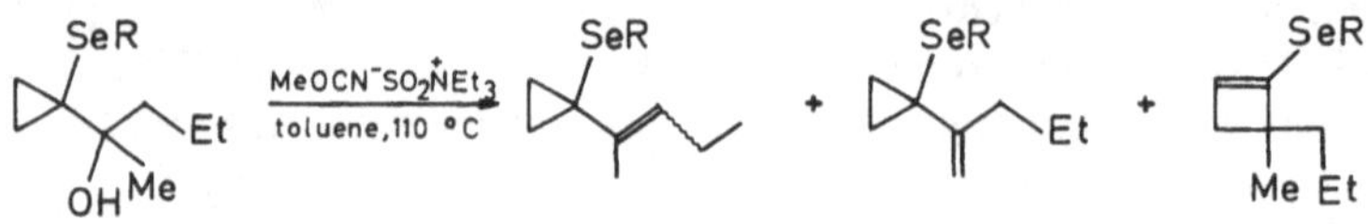

R	time (h)	overall yield	ratio				
Ph	2.5	77 %	49	:	27	:	24
Me	38	65%	61	:	35	:	4

Scheme 53

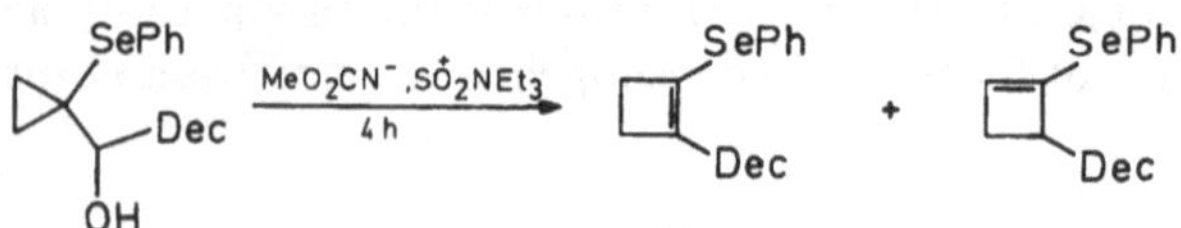

	ratio		
60% overall	85	:	15

Scheme 54

and 1-(1-phenylthio) cyclopropyl carbinols, respectively, is applicable only [37)4] to those 1-(1-seleno) cyclopropyl carbinols whose hydroxyl group is attached to a fully substituted carbon atom (Schemes 52, 53),. In other cases it produces[37] 1-seleno-1-cyclobutenes instead (Scheme 54). Even so, the corresponding selenocyclobutene is always formed in addition to the vinylcyclopropane (Schemes 52, 53). It accounts[37] for only a few percent in the methylseleno series but can be up to $\sim 20\%$ in the phenylseleno cases. Moreover, in most cases the reaction does not lead to one unique vinylcylopropane but instead to a mixture of all the possible regio- and stereoisomers in which the one possessing the more substituted carbon-carbon bond predominates (Scheme 53, 54).

4.2.1.2 *Synthesis of 1-Silyl-1-Vinylcyclopropanes*

Dehydration of -1-(-1-silyl)cyclopropyl carbinols — was disclosed by Paquette [82,

[4] For a similar observation with phenythio analogues see ref. [131]

[138,182] and later by Cohen [81]. Paquette observed (as we did in the seleno series [37]) a large difference of reactivity between those compounds possessing a hydroxyl group attached to a fully substituted carbon atom (Scheme 55) [82] or not (Scheme 56a) [138].

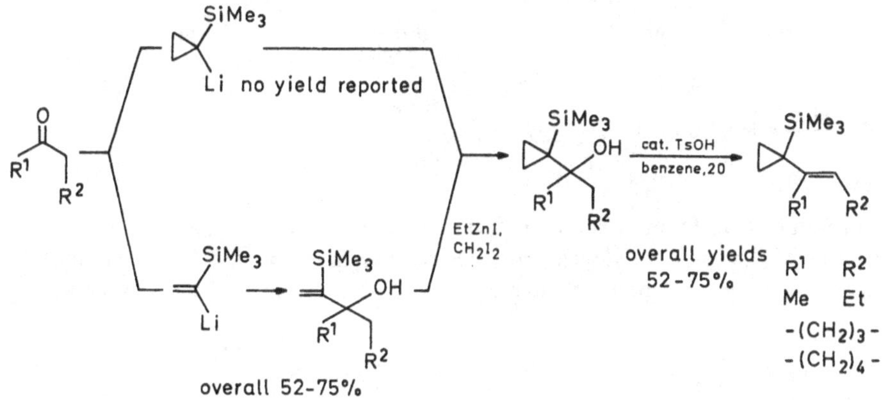

Scheme 55

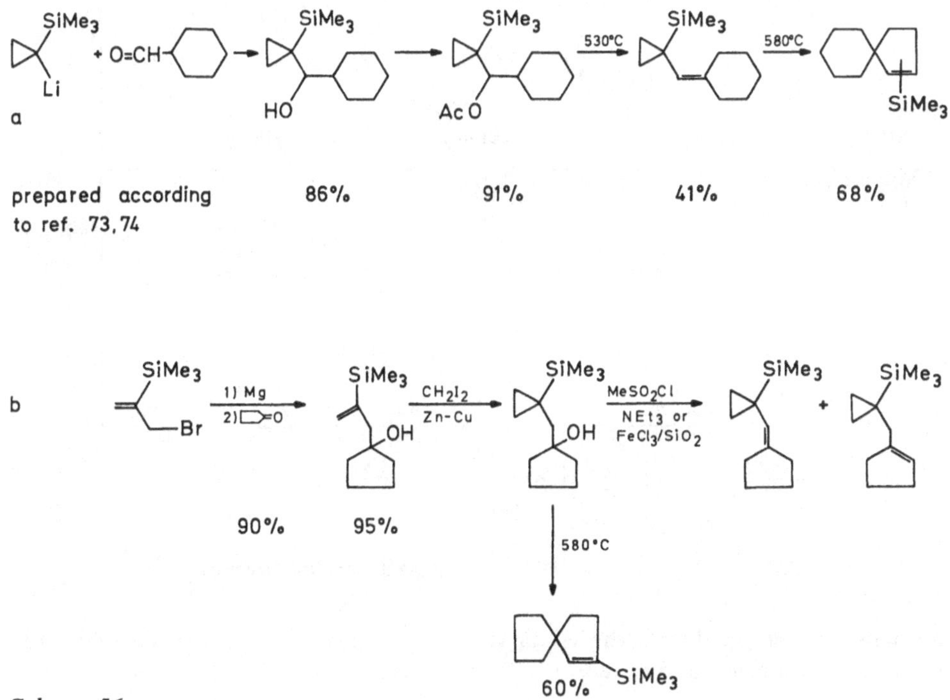

Scheme 56

Vinylcyclopropanes are formed in good yield in the first case on reaction with catalytic amounts of p-toluenesulfonic acid (TsOH), whereas the elimination requires [138] heating of the corresponding acetate up to 530 °C in the other cases (Scheme 56a). Of course, the presence of an activating group [182], such as a keto group (Scheme 57)

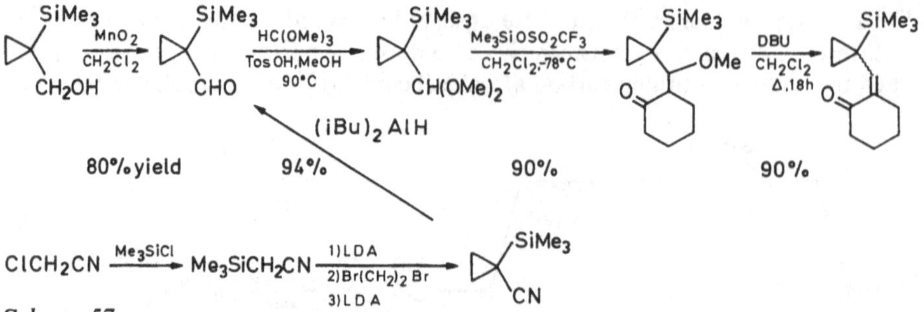

Scheme 57

in a suitable position favours the elimination reaction. Paquette has also described the synthesis of the same vinylcyclopropanes by dehydration [138, 182] of 1-(1-silyl)-1-(2′-hydroxyalkyl) cyclopropanes (Schemes 56b, 58a), but, except in the cases where

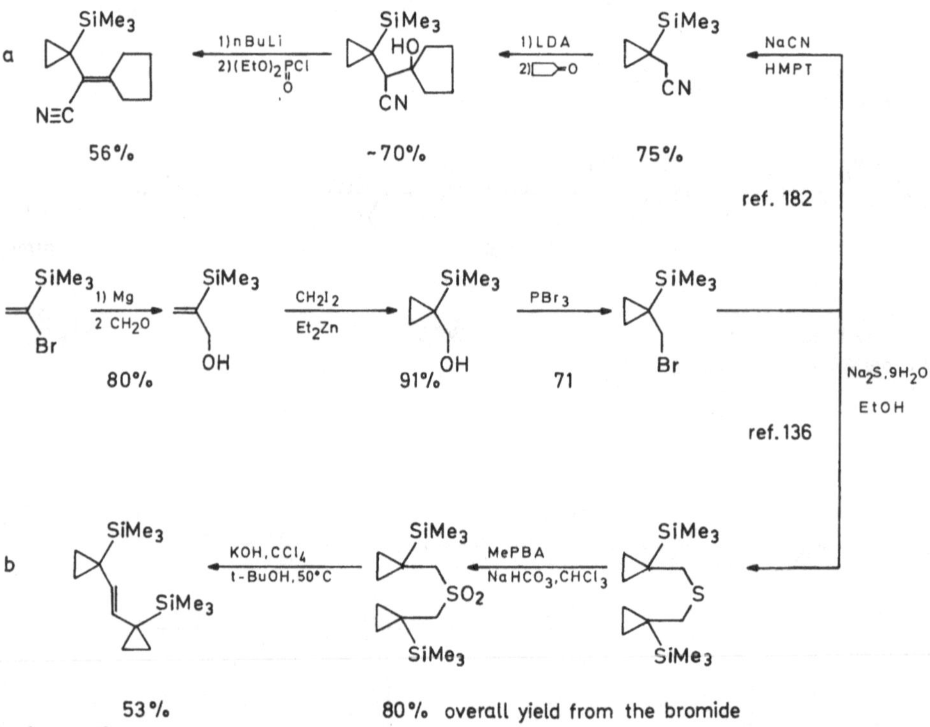

Scheme 58

an activating group directs the elimination [182] (Scheme 57), a mixture of all the possible regiosomeric and stereosomeric olefins is formed [138] (Scheme 58b).

4.2.2 Synthesis of 1-Hetero-1-vinylcyclopropanes by Elimination of Two Heteroatomic Moieties

This strategy (Scheme 59) has been used with equal success in the seleno series[31] (Schemes 45, 59–61) as well as in the thio[35, 45] (Scheme 62), the bromo[35], (Scheme 62) and the silyl[77, 138] (Schemes 63, 64) series. All The transformations involve the

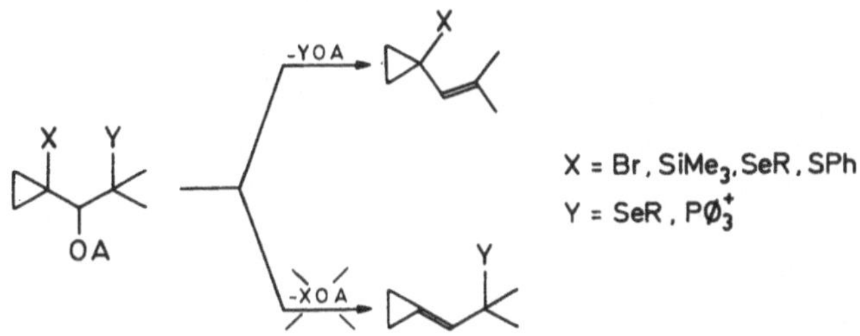

Scheme 59

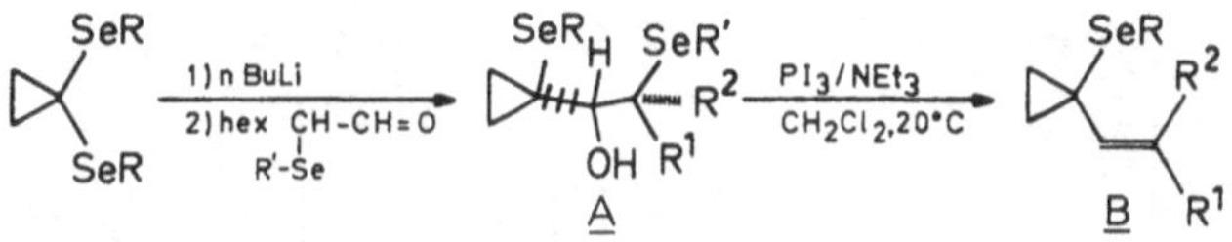

R	R'	R_1	R_2	yield in A	yield in B $\left(\frac{E}{Z + E}\right)$
Me	Me	H	hex	71%	93% (98)
Ph	Me	H	hex	65%	86% (94)
Me	Me	Me	hex	81%	91% (85)*

* the stereochemical assignement has not yet been performed

Scheme 60

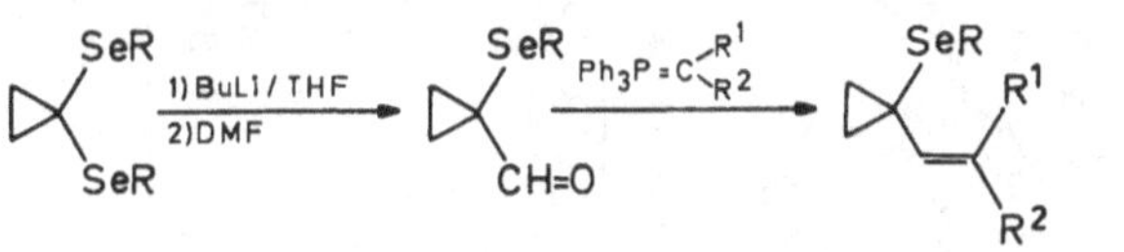

R	yield	R_1	R_2	yield (Z/Z + E)
Me	84%	hex	H	93% (98)
Ph	74%	hex	H	86% (94)
Me	84%	hex	Me	91% (85)*

* The stereoisomericassignement has not yet been done

Scheme 61

formation of the desired double bond by β elimination reaction on a suitable substrate of hydroxide or an alkoxide group and of a selenyl [31,35,77,138] (Schemes 59, 60, 62–64) or an alkoxide group and a phosphonium moiety [31,35,77] (Schemes 61–63). or two alkoxide groups [138] (Scheme 65).

The starting materials have been prepared in good to high yields from α-hetero-substituted cyclopyllithiums and α-selenoaldehydes [31,35,77] (Schemes 60, 62, 63), or in a more convergent manner from 1-formyl-heterosubstituted cyclopropanes and phosphonium ylides [31,35,45,77] (Schemes 61–63) or α-seleno alkyllithiums [31,138] (Schemes 59, 64).

The last approach is the least stereoselective [31] among the three mentioned. It is, however, interesting to note that 1-seleno-1-seleno-1-vinyl cyclopropanes are formed

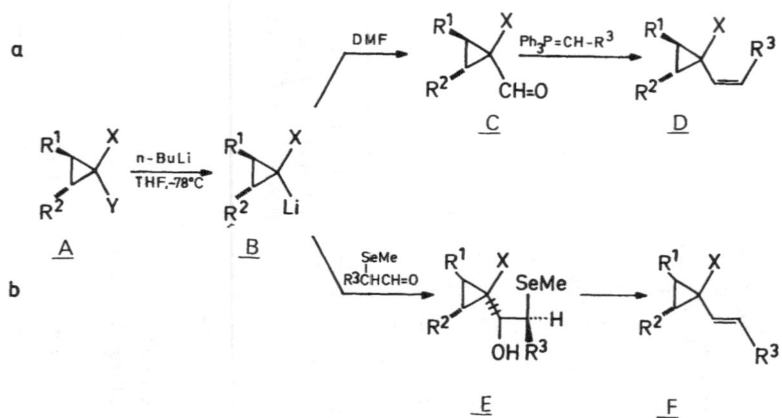

A and B	R_1	R_2	X	Y
	H	H	SPh	H
	Me	Me	Br	Br

C	X	yield	D	X	R_1	R_2	yield	(Z/Z + E)	Ref.
	SPh	45%		SPh	H	H	hex	96%	45)
	Br	56%		Br	Me	Me	Bu	50%	35)

E	X	yield	F	X	R_1	R_2	R_3	yield (E/z + E)	Ref.
	SPh	72%		SPh	H	H	hex	80%	45)
	Br	73%		Br	Me	Me	Bu	72%	35)

Scheme 62

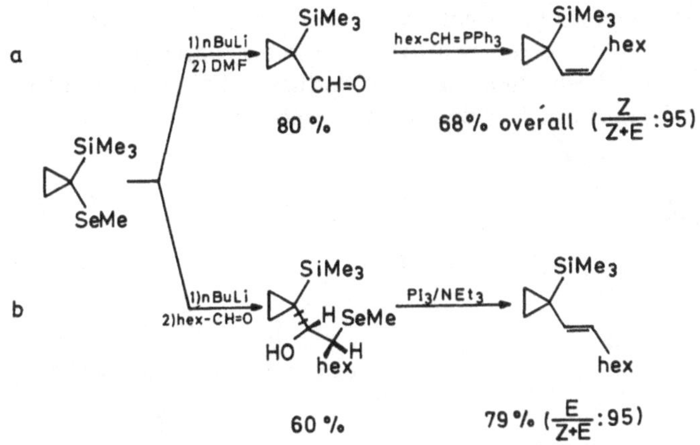

Scheme 63

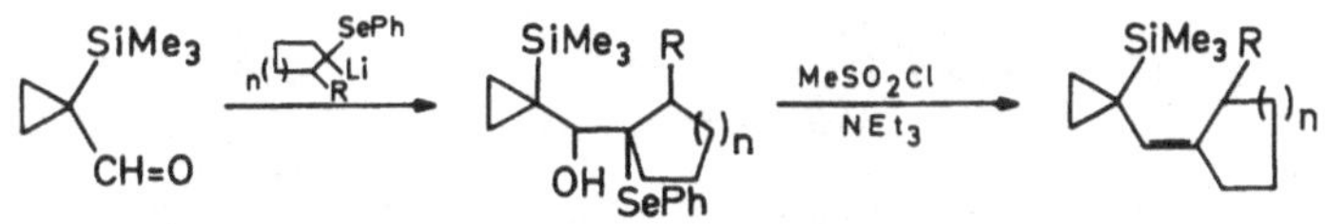

n	R	yield	yield
1	H	81%	80%
2	H	88%	78%
2	Me	83%	75%

Scheme 64

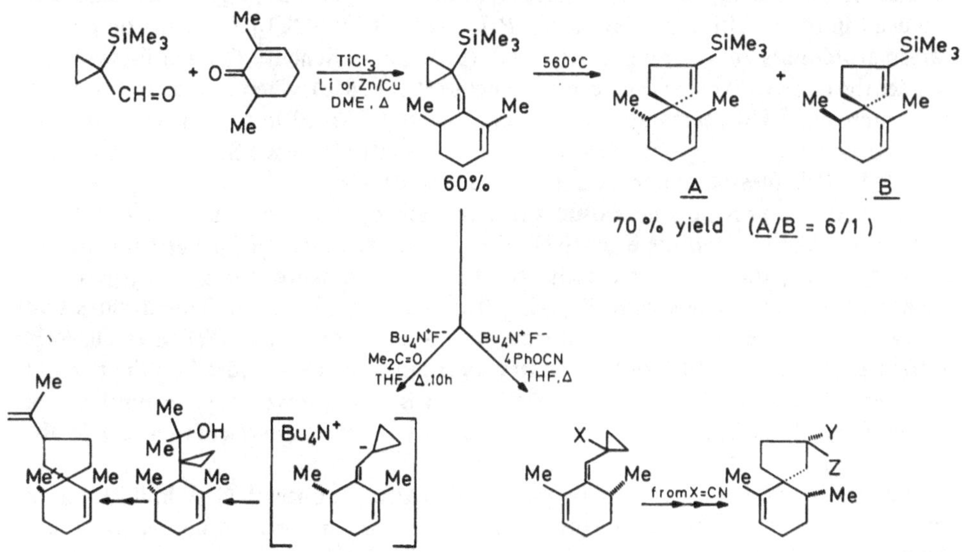

x=CN 69% z=H y=Me_2C–OH hinesol
x=H 20% zv x = =C $(Me)_2$ β-vetivone

Scheme 65

rather that alkylidene cyclopropanes whatever groups is attached to the two selenium atoms [31,35] on 1-(1-seleno)-1(1′-hydroxy-2′-selenoalkyl)cyclopropanes (Schemes 59, 60). This method has been extended to the synthesis of vinylcyclopropanes in which the vinyl group is substituted by one or two heteroatomic moieties [185] (Scheme 66).

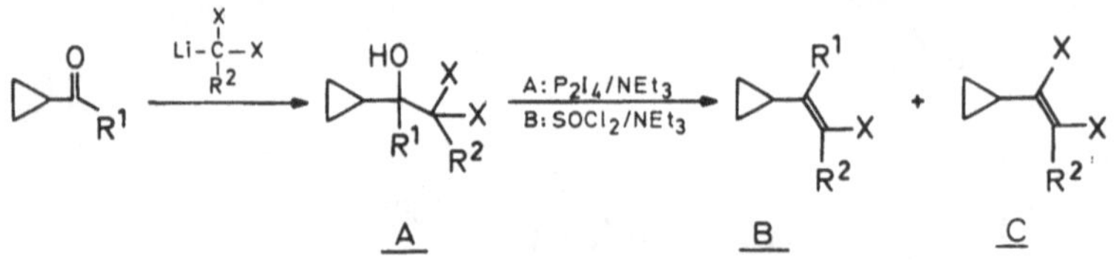

	R_1	X	R_2	yield in A %	method	yield in (%) B		C
a	H	SMe	H	76%	A	53	:	16
b					B	trace	:	50
c	H	SeMe	H	46%	A	65		—
d	H	SMe	SMe	66%	A	31	:	28
e					B	80	:	90
f	H	SeMe	SeMe	62%	A	85	:	—
g					A	66	:	—
d	Me	SMe	SMe	83%	B	69	:	—

Scheme 66

It uses cyclopropyl aldehydes or cyclopropyl methyl ketones and α-lithiothio- and α-lithioselenoacetals or α-lithiothio- or α-lithioselenoorthoesters as the starting materials. The elimination of the hydroxyl and of the sulfenyl or selenenyl moiety has been achieved in the resulting product using P_2I_4/NEt_3 or by $SOCl_2/NEt_3$. In some cases variable amounts of rearranged products (product *C* Scheme 66) are formed [185], besides the expected ketene thioacetals (Scheme 66a, b, d, e), when an aldehyde is used. It is interesting that rearranged compounds are not formed in the seleno series and that their amount in the sulfur series increases dramatically when $SOCl_2$ is used rather than P_2I_4 (Schemes 66, compare b to a and e to d).

The method involving α-heterosubstituted cyclopropyllithiums and α-selenenylaldehydes provides[31,35,77] quite exclusively the (E) stereoisomers of 1-heterosubstituted-1-vinylcyclopropanes after treatment of the 1-(1-heterosubstituted)-1′-hydroxy-2′-selenoalkyl)cyclopropanes with PI_3/NEt_3 (Schemes 60, 62b, 63b). This implies that the two steps involved in the transformations are stereoselective (Schemes 60, 62b, 63b). The stereochemical result of the first step can be rationalized[123] on the basis of the Cram-Karabatsos-Felkin rules[124,125] and it is well-known that the synthesis of olefins from (β-hydroxyalkyl)selenides occurs by a formal *anti* elimination of the hydroxyl and selenyl moieties [4–9,11,12,123,163].

The reaction which uses the phosphorus ylide also produces[31,35,77] highly stereoselectively 1-heterosubstituted-1-vinylcyclopropanes, but now the (Z)stereoisomers are the major ones (Schemes 61, 62a, 63a). It is well-known that in the Wittig reaction the betaines which lead to the (Z)-disubstituted olefins are formed under kinetically controlled conditions and that any feature which favours the elimination of triphenylphosphine oxide from these betaines should increase the amount of the (Z)stereoisomer. The possible stabilization of a β-carbenium ion by the cyclopropane ring and the heteroatomic moiety could explain the results observed.

This reaction has been successfully applied[117] to the synthesis of 1-seleno-1-vinyl-cyclobutanes from α-selenocyclobutyl carboxaldehyde (Scheme 67) and various

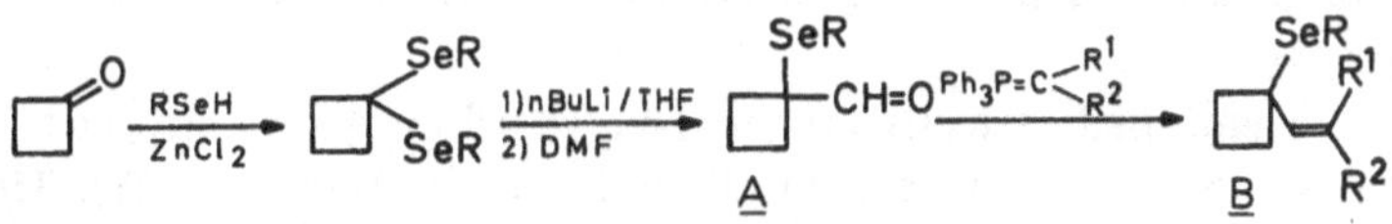

A:	R	yield	B	R	R_1	R_2	yield
	Me	62		Ph	H	H	92%
	Ph	77		Me	Et	H	75%
				Ph	Et	H	86%
				Me	hex	H	80%
				Ph	hex	H	89%
				Me	Ph	H	75%
				Ph	Ph	H	77%
				Me	Me	Me	0%
				Me	Me	Et	0%

Scheme 67

phosphorus ylides bearing at least one hydrogen on the carbanionic center. However, it does not work[117] with those homologs bearing two alkyl substituents there (Scheme 67). Unfortunately the stereochemistries of the vinylcyclobutanes prepared by this method have not yet been determined. An interesting observation was made with the parent phenylseleno derivative during this work[117], since it was found (Scheme 68) that this compound has a high propensity to rearrange[25] to the regioisomeric

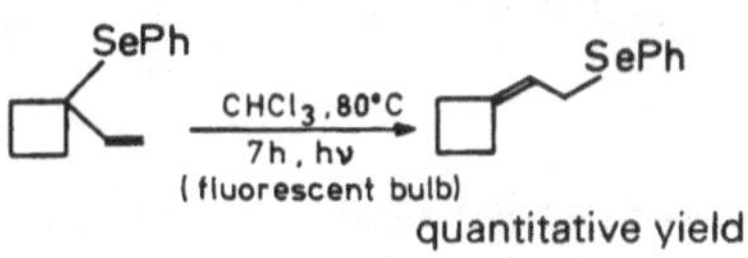

Scheme 68

alkylidene cyclobutane on irradiation with UV light or light from a fluorescent bulb. This isomerisation takes place [25] with other allylselenides bearing a terminal olefinic group but not with the cyclopropyl analogs, probably due to the strain involved.

4.2.3 Miscellaneous Syntheses of 1-Hetero-1-vinylcyclopropanes

Other syntheses of 1-heterosubstituted vinylcyclopropanes have been proposed, especially in the silyl series. They involve the Ramberg-Backlund methodology[136, 186] (Scheme 57b) or the McMurry strategy [136, 138, 187, 188] (Schemes 65, 69).

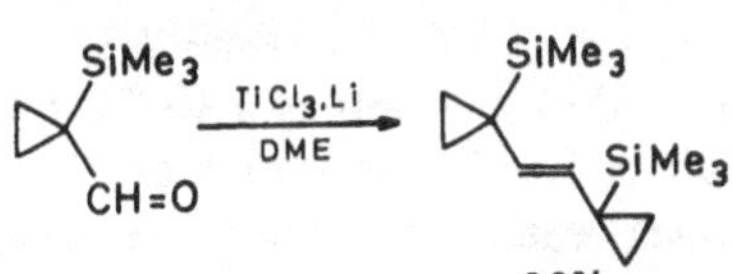

38%

Scheme 69

4.3 Synthetic Transformations Involving 1-Heterosubstituted-1-Vinylcyclopropanes

4.3.1 Reactions Involving 1-Seleno-1-vinylcyclopropanes

4.3.1.1 Synthesis of 1-Functionalized-1-Vinylcyclopropanes Via 1-Lithio-1-vinylcyclopropanes

As already mentioned, 1-methylseleno-1-vinylcyclopropanes produce[36] 1-lithio-1-vinylcyclopropanes on reaction with alkyllithiums (Schemes 17, 18). These organometallics react with various electrophiles, such as water, alkylhalides, and carbon monoxide, to produce the corresponding 1-functionalized vinylcyclopropanes (Scheme 17) which retain both the regio- and stereochemistry (with the exclusion of the styryl derivative [36] (Scheme 18)) originally present in the starting vinylselenide [36]. A mixture of functionalized alkylidene cyclopropanes and vinylcyclopropanes are formed if acetaldehyde or bezyldehyde are reacted [36] instead (Scheme 70). The higher

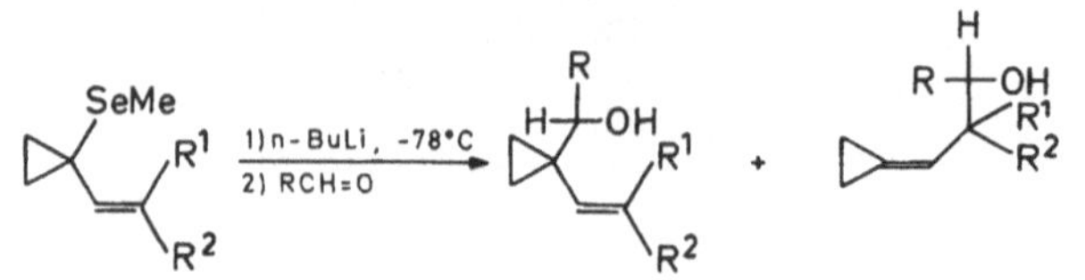

A B C

R_1	R_2	R	conditions	yield			
hex	H	Me	TMF	55%	50	:	50
H	hex	Me	TMF	68%	80	:	20
hex	H	Ph	TMF	90%	25	:	75
hex	H	Ph	TMF-TMEDA	88%	20	:	80
H	hex	Ph	TMF	96%	46	:	54

Scheme 70

percentage of functionalized vinylcyclopropane is found when starting with the (E) stereoisomers. Interestingly the functionalized cyclopropanes retain the stereochemistry present on the starting vinyl selenides (Schem° 70). Similar derivatives have been obtained [138] on reaction of a mixture of 1-silyl-1-vinylcyclopropanes, acetone, and tetrabutylammonium fluoride (Scheme 65), see Sect. 4.3.2.1.

4.3.1.2 Syntheses of Functionalized Alkylidene Cyclopropanes

α-Selenovinylcyclopropanes have been readily transformed[45] to functionalized alkylidene cyclopropanes by a sequence of reactions which involve as the key step the transposition of the corresponding allylic selenoxides or selenonium ylides (Scheme 71).

By a Reaction Involving the Rearrangement of an Allylic selenoxide

The transposition of allylic selenoxides to allyl alcohols is a well-established reaction[24–27, 189] which parallels the one described with sulfur analogs[190]. It was found, however,[45] that phenylselenoxy derivatives rearrange much more easily than their phenylthio or their methylseleno analogs.

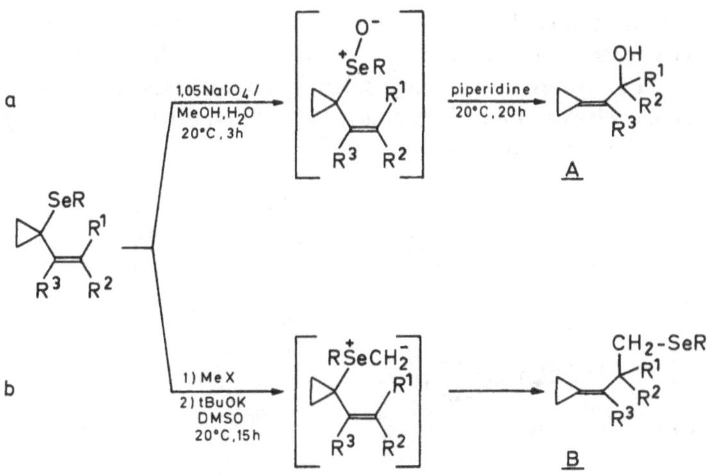

A	R	R_1	R_2	R_3	overall yield (%)
1	Ph	H	hex	H	75
2	Ph	H	Ph	H	70
3	Ph	oct	oct	H	75
4	Ph	$(CH_2)_3$		hex	92
5	Ph	H	pent	hex	92
6	Me	H	pent	H	16

B	Alkylation conditions	R	R_1	R_2	R_3	overall yield (%)
1	$MeSO_3$/ether	Me	H	hex	H	85
2	MeI-$AgBF_4$	Me	H	hex	H	53
3	MeI-$AgBF_4$	Ph	H	H	H	60
4	MeI-$AgBF_4$	Me	Me	hex	H	85
6	$MeSO_3F/CH_2Ch$	Ph	H	pent	hex	48*

* as a 85/15 mixture of stereoisomers

Scheme 71

Those belonging to the cyclopropyl series, as expected, were found [45,184] to be less prone to rearrange since the formation of the seleninate, the postulated intermediate, is disfavored due to the strain introduced during the concomitant formation of the alkylidene cyclopropane. For example, allyl alcohols are formed in only 30% yield if the vinylcyclopropyl selenides are reacted at room temperature with an aqueous solution of hydrogen peroxide in THF, under conditions which almost quantitatively produce allyl alcohols from straight-chain phenyl or methylseleno analogs [45]. The best results have been obtained if the two different steps required for such reactions are separated. 1-(Phenylseleno)- and 1-(methylseleno)-1-vinylcyclopropanes have been quantitatively oxidized to the corresponding selenoxides [45] (Scheme 71a) and the rearrangement to the allyl alcohol has been forced by adding piperidine to the selenoxide dissolved in a non polar solvent such as cyclohexane. Piperidine plays the role of a selenophilic agent able to destroy the seleninate and therefore to shift the equilibrium towards the formation of the allyl alcohol. In fact N-(phenylseleno)piperidine has been isolated at the end of the reaction involving phenylseleno derivatives [35]. Best results were observed[45] with phenylselenoxy derivatives which produce (in cyclohexane and in the presence of piperidine after a few hours at 20 °C) the allyl alcohols in very good yields (Scheme 71a, entries 1–5). Methyl selenoxy compounds react slowly [45] (Scheme 71a, entry 6), whereas phenyl [45] and methyl [184] sulfoxy analogs remain inert, even under drastic conditions (Scheme 72a) [44].

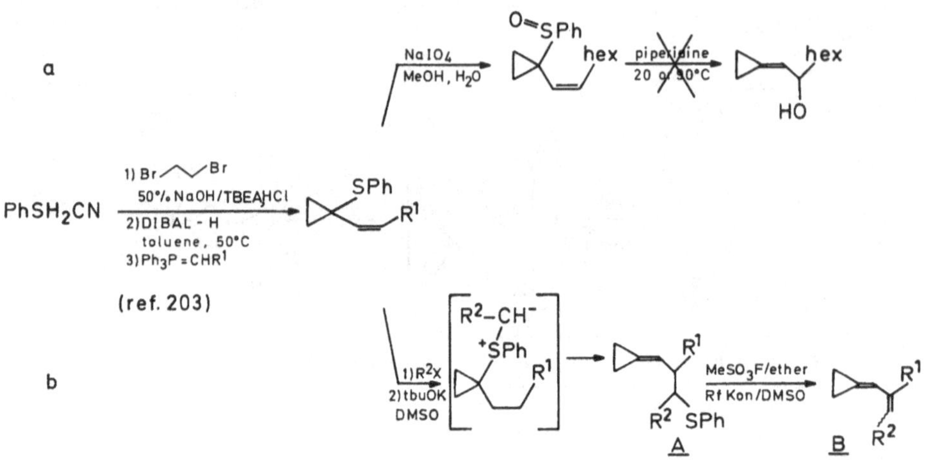

alkylation conditions	R_1	R_2	A overall yield	B overall yield
$MeI\text{-}AgBF_4/CH_2Cl_2$	hex	Me	77%	—
$EtI\text{-}AgBF_4/CH_2Cl_2$	hex	Et	71%	—
$MeSO_3F$/ether	H	Me	83%	70%
$MeSO_4$	H	Me		(ref 35)

Scheme 72

By a Reaction Involving the Rearrangement of an Allylic selenonium ylide

Vinylcyclopropanes bearing a selenenyl or a sulfenyl moiety on the cyclopropane ring have been readily alkylated[45] with magic methyl in ether or with alkyliodide/silver tetrafluoroborate in dichloromethane (Schemes 71b, 72b). Further treatment of the resulting selenonium/sulfonium salts with potassium tert-butoxide in DMSO at 20 °C for 15 hrs provides an efficient synthesis [45] of selenides or sulfides bearing a cyclopropylidene moiety in the β-position (Schemes 71, 72), which results from the 1,3-transposition of the ylide intermediately formed. These selenides or sulfides can again be transformed [45] to the corresponding salts but, now, on further treatment with potassium tert-butoxide in DMSO, the ylides formed lead selectively to allylidene cyclopropanes via a regioselective elimination reaction of the type used for the synthesis of alkylidene cyclopropanes and already described in Sect. 4.1.1.1. (Schemes 72, 73).

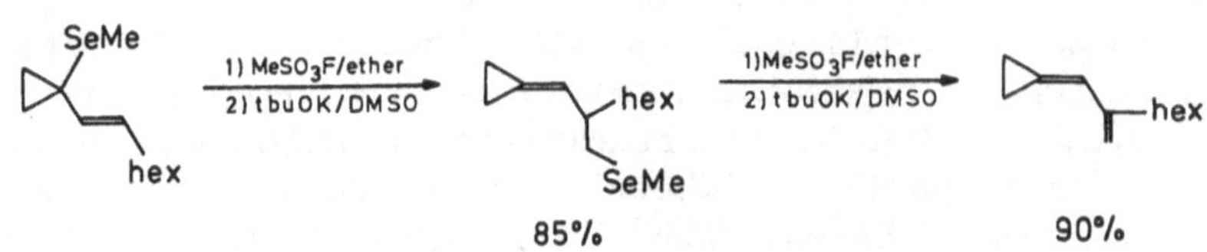

Scheme 73

4.3.1.3 Synthesis of Carbonyl compounds

Vinyl cyclopropanes have been subjected [170] sequentially to the reaction of ozone and of P_2I_4 or PI_3. Depending on the conditions used, an α-selenocyclopropyl ketone (1 mol eq P_2I_4, −78 °C, 0,75 hr) or a γ-iodo ketone (2 mol eq P_2I_4, −78 °C to 20 °C, 1 hr) are formed in high yield (Scheme 74). The activity of the phosphorus reagents is

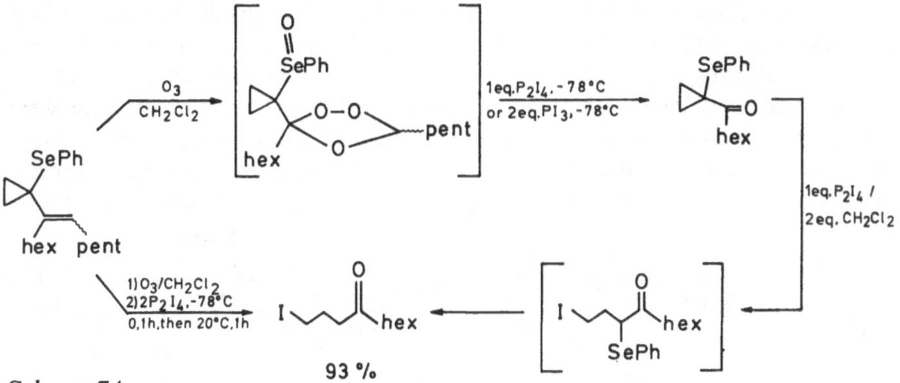

Scheme 74

particularly impressive since they perform in a stepwise manner the reduction of an ozonide, the reduction of a selenoxide, the ring opening of an activated cyclopropane, and the reduction of an α-seleno ketone (Scheme 81). Finally 1-seleno-1-vinylcyclopropanes have been found, as will be shown in the next section, to be valuable precursors of cyclobutanones [35,36] (Scheme 75).

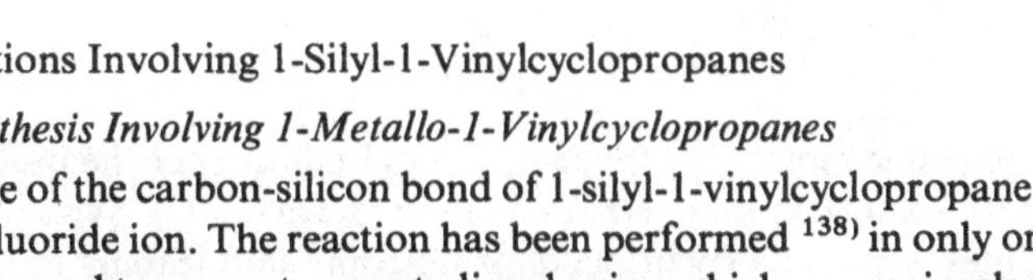

Scheme 75

4.3.2 Reactions Involving 1-Silyl-1-Vinylcyclopropanes

4.3.2.1 Synthesis Involving 1-Metallo-1-Vinylcyclopropanes

The cleavage of the carbon-silicon bond of 1-silyl-1-vinylcyclopropane has been achieved by the fluoride ion. The reaction has been performed [138] in only one case (Scheme 65) but has severd to generate a pentadienyl anion which was regioselectively hydroxyalkylated on the cyclopropane ring [138] (Scheme 65). This reaction has been successfully used for the synthesis of (±) α-vetispirene [138] (Scheme 65).

Fluoride-induced desilylation-cyanation studies have been undertaken [138] with several electrophilic cyanogen type reagents. The best results were attained [138] when the 1-silyl-1-vinylcyclopropane shown in Scheme 65 was heated in a THF solution containing 4 equivalents of phenylcyanate. This reaction has been used [138] as a key step for the synthesis of (±)-hinesol and of (±)-β-vetivone (Scheme 65).

4.3.2.2 *Thermal Rearrangement to Cyclopentene Derivatives*

Vinylcyclopropanes are also suitable starting materials for the construction of five-membered cycles. The vinylcyclopropane rearrangement first discovered in the late 1950s [191,192] immediately became the subject of intense mechanistic investigations [173,193,194] and was used for the cyclopentene anelation [165,171,195,197] or cycloheptane anelation [198–200]. This reaction was successfully applied to vinylcyclopropanes bearing a silyl moiety by Paquette [137,138] and was found particularly useful for the synthesis of vinylsilanes as part of a five-membered cycle. These have been in some cases subjected to electrophilic substitution [137,138,201,202] (Schemes 58b, 65, 82–84). The ring expansion takes place when the thermolysis is conducted between 570° and 670 °C and only occurs with those compounds which do not possess [137] steric inhibition around the carbon-carbon bond (Scheme 76, compare a and e). Incorporation of the double bond (Scheme 76, compare a and e). Incorporation of the double

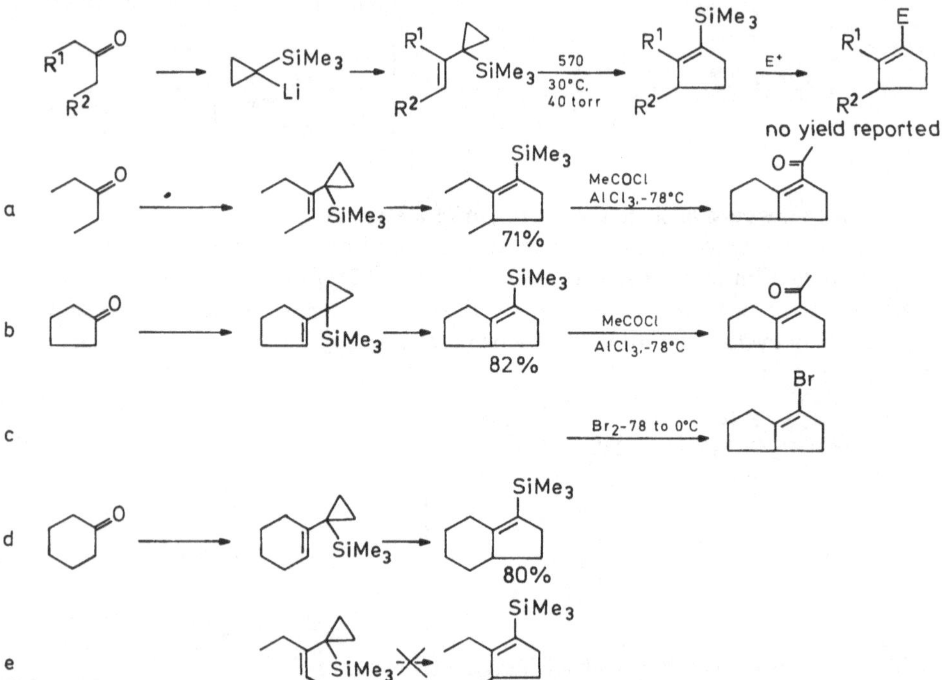

Scheme 76

bond into a cyclic enone moiety [137] (Scheme 77) did not disrupt the bond relocation process, although a somewhat more elevated temperature (660 °C) was required. The resulting compound is particularly prone to prototropic shift leading to an allylsilane in which desilylation occurs readily during the chromatographic purification [137] (Scheme 77). The presence of a trimethylsilyl group at C1 of a vinylcyclopropane has

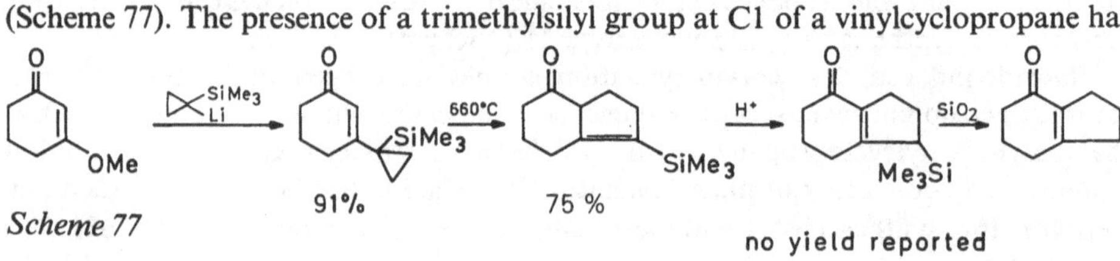

Scheme 77

been shown to have important kinetic consequences [137]. The electropositive character of this substituent retards the bond reorganization and decreases the isomerisation rate [137] (Scheme 78). The extent to which such isomerization is retarded by the 1-trimethylsilyl group has been estimated by the increase in activation energy of the thermal rearrangement of those vinylcyclopropanes shown in Scheme 55. This rearrangement was found [137] to require 10 kcal/mol more than for the corresponding vinylcyclopropane which does not bear the trimethylsilyl group.

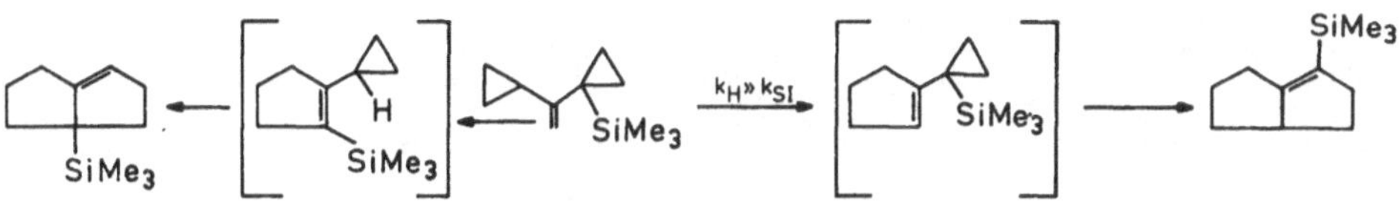

B A

Scheme 78

4.4 Diels-Alder Reactions Involving Allylidene Cyclopropanes

Allylidene cyclopropanes proved to be particularly reactive dienes especially if used as one of the partners in the Diels-Alder reaction [77,107,136,203–205]. They differ in that respect from other dienes bearing two alkyl groups at one terminus, which are inert in a [2 + 4]cycloaddition even with very reactive dienophiles and often lead to products resulting from their rearrangement [107,206,213]. Most of the reactions have been performed on the parent diene [107,214] (Schemes 79–81). although a few deal with analogs bearing alkyl groups on the olefinic moiety and/or on the cyclopropane ring [77] (scheme 82) or with dicyclopropylidene ethane [136] (Scheme 83). The latter compounds is much less reactive than other members of the series, but even so, its reactivity is exceptional despite the obvious congestion at its bonding centres [136]. Functionalized spiro[2,5] octenes are formed in good yield with highly electrophilic dienophiles such as tetracyano ethylene [107,136] dialkyl acetylenedicarboxylates [107,136], p-quinones [77,107,214], maleic and fumaric diesters [107], maleic anhydride [77,107], maleimide [136], azodicarboxylate [203] and N-methyltriazolinedione [136] (Schemes 79–83). Olefins mono-activated by an alkoxycarbonyl, a nitrile, an aldehyde or a keto group also lead to the corresponding spiro[2,5]octenes [107] (Scheme 81) but the reaction is much slower and the yields are lower [107]. The presence of an alkyl substituent on the dienophile dramatically lowers the yields if it is branched at the α-position and almost inhibits the reaction if it is located at the β position of the activating group. In these specific cases the temperature at which the reactions are performed was found [107] to be critical. The best compromise was found to be 100 °C since spiro-octene derivatives are formed in much lower yields at lower (75 °C) or at higher (150 °C) temperatures. In the latter case the dimerisation of the diene was observed, especially with nonreactive dienophiles. While these reactions show the stereospecificity usually observed in the Diels-Alder reaction (see Scheme 79e, f), their regioselectivity is significantly different from the one usually reported for 1-substituted dienes which are known to give predominantly the ortho-type adduct [215]. In the case of allylidene cyclopropanes a reversed tendency was found and the meta-type adduct was mainly formed [102] (Scheme 81). The regioselectivity is high in the case of α,β-unsaturated esters, ketones, and aldehydes and somewhat less pronounced with nitroethylene.

The formation of the meta-type adducts instead of the ortho ones can be rationalized on the basis of a mechanism involving radicals or ionic intermediates in the transition state (scheme 84). Intermediate structures (*1b*) and (*2b*) can account for the results usually observed with dienes bearing alkyl groups whereas intermediates such as (*1*a) and (*2*a), in which the strain due to the cyclopropyl ring is

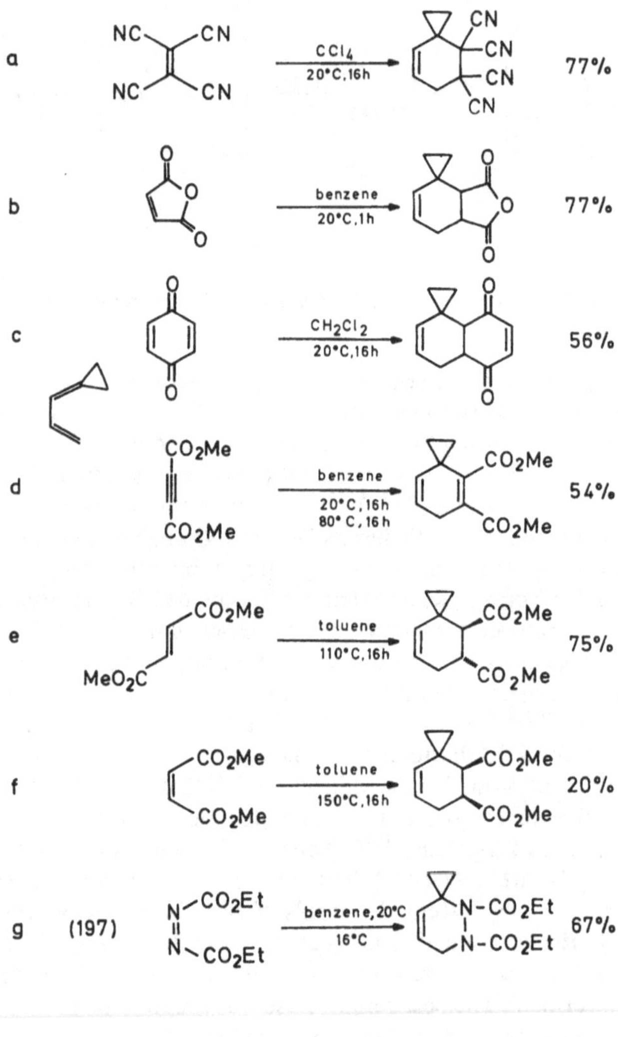

Scheme 79

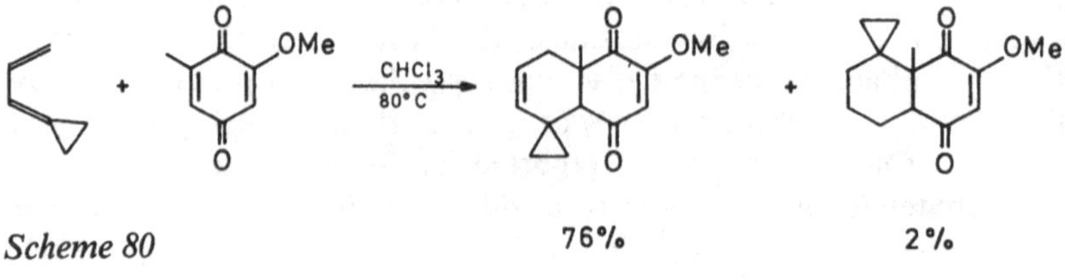

Scheme 80

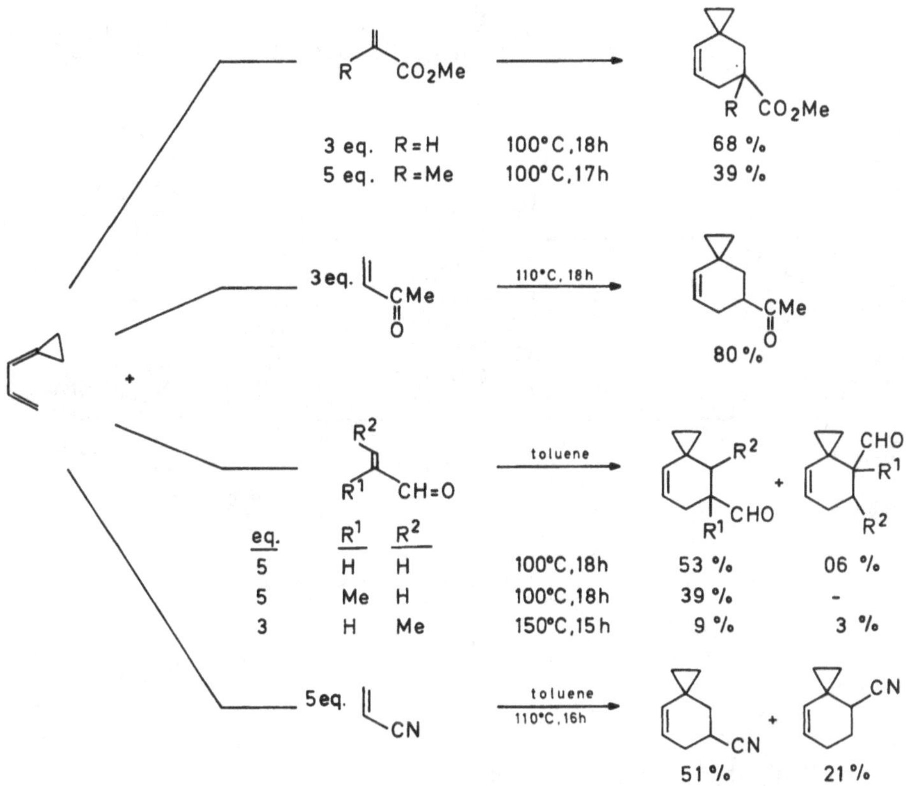

* The structure of the regioisomers has not been assyned

Scheme 81

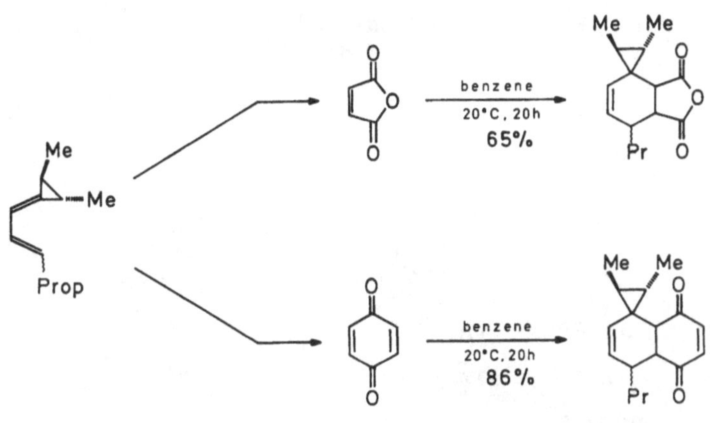

Scheme 82

can explain the results observed in the case of cyclopropyl derivatives. This would be the reason for the unusually high reactivity of such dialkylated dienes. The vinylcyclopropane system present in the spirooctene derivatives can serve as a precursor of spirooctanes, of gem-dimethylcyclohexanes or of ethylcyclohexanes, as has been shown [107] in the case of the acrylate adduct (Scheme 85). The carbon-carbon double bond present in the cyclohexene can be selectively reduced [102] (without any harm to the

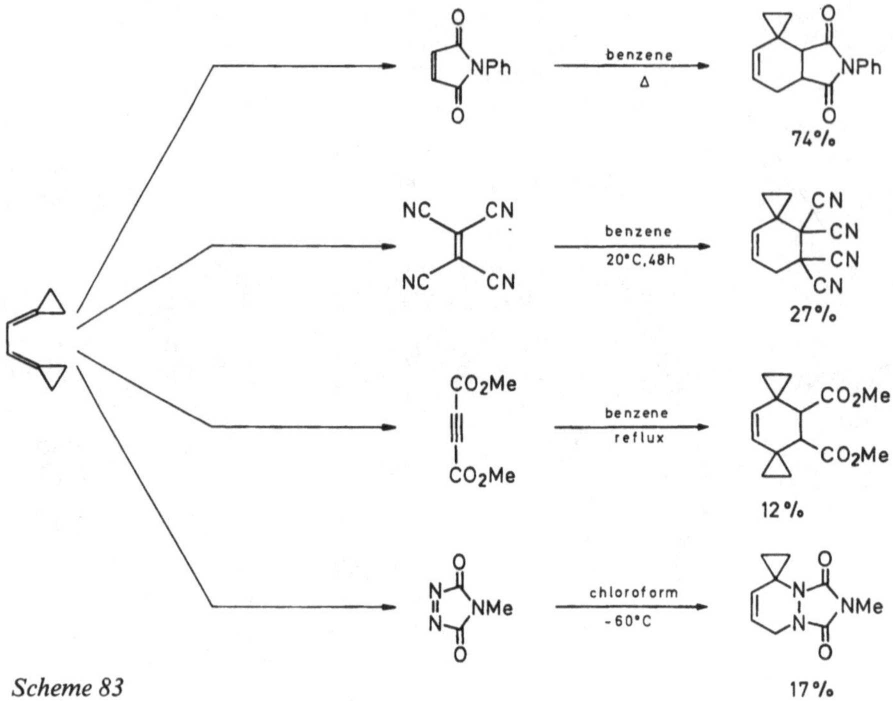

Scheme 83

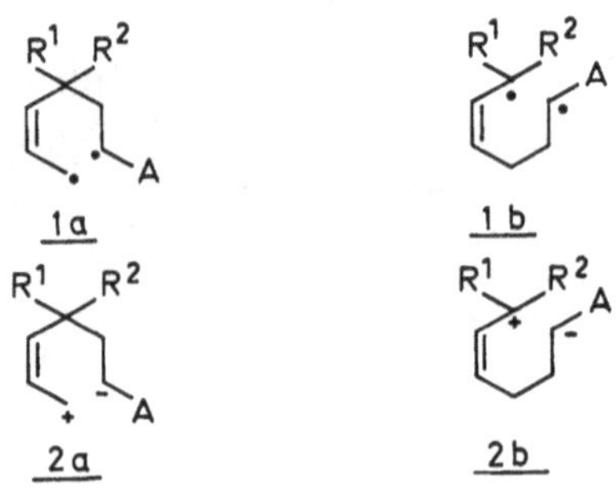

Scheme 84

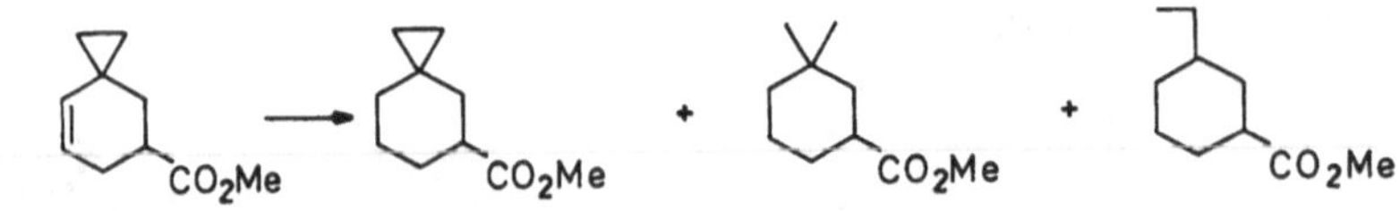

RhCl : PPh_3	84%	100	–	–
Pt / AcOH	93%	-	97	3
10% Pd/C, THF, H_2O	92%	16	–	84

Scheme 85

cyclopropane ring) with hydrogen in the presence of the Wilkinson catalyst [$(Ph_3P)_3RhCl$. 0.1 eq in benzene 20 ëC, 10 hr]. On the other hand, both the carbon-carbon double bond and the cyclopropane ring can be reduced [107] if platinum is used (Pt, 0.3 eq EtOH, 20 °C, 4 hr, 93%), while methyl 3-ethylcyclohexanecarboxylate formally formed [107] through a 1,5-addition of hydrogen to the vinylcyclopropane is obtained when the reaction is conducted over palladium [Pd/C 10%, 20% (w/w) THF/H_2O (1:1), 20 ëC, 4 hr, 92% yield].

4.5 Syntheses of Carbonyl Compounds by Ring-Enlargement Reactions

One of the interesting features of functionalized cyclopropyl and cyclobutyl derivatives is, without doubt, their ability to release the strain and to produce larger rings or straight-chain derivatives. This has been demonstrated by the pioneering work of Julia [166] on cyclopropyl carbinols, which in acidic media lead to homoallyl alcohols [166, 167, 216], and by the well-known propensity of β-heterosubstituted (RO [85, 217], RS [71, 131, 132], RSe [87, 135], Br [71, 73], $RSeO_2$ [67]) such as cyclopropylcarbinols and oxaspiropentanes [61, 63, 64, 67] to produce cyclobutanones in acidic [61, 63, 64, 67, 71, 81, 92, 132, 133] or basic media [67, 71, 73] (Scheme 86). But while many β-heterosubstituted

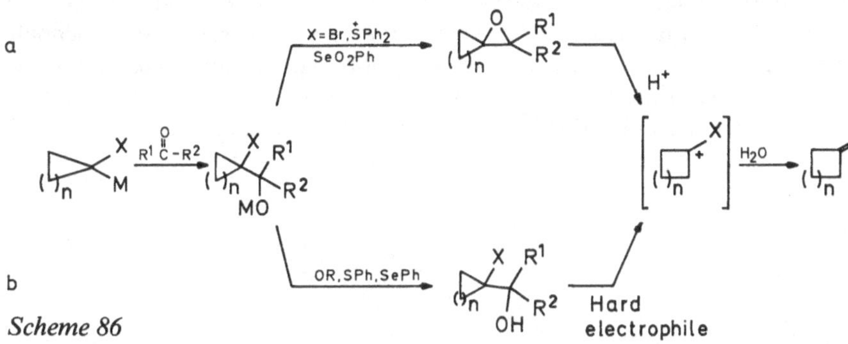

Scheme 86

cyclopropylcarbinols (RO [85, 217], RS [71, 131, 132], RSe [87, 135]) lead directly to cyclobutyl derivatives when the hydroxyl group is transformed to a better leaving group. 1-(1-silyl)cyclopropylcarbinols lead instead to 1-silyl-1-vinylcyclopropanes [82] (Scheme 87) when reacted with catalytic amounts of p-toluene sulfonic acid in benzene at 20 °C and to isomerically pure cyclopentylidene tosylates (with an exocyclic double bond) or

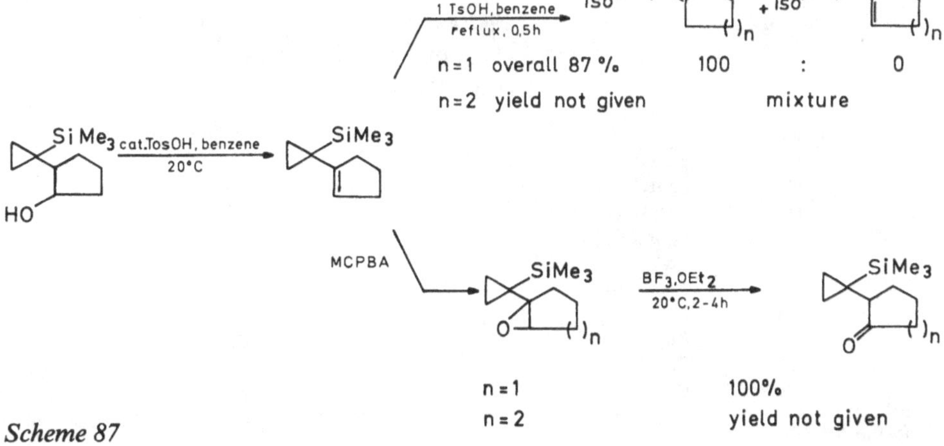

Scheme 87

to a mixture of cyclohexylidene and cyclohexenyl tosylates by the ring opening of the cyclopropyl moiety on reaction [82] with 1 equivalent of p-toluenesulfonic acid in refluxing benzene (Scheme 87a). The energetic cost of positioning a positive charge α to silicon is not outweighed by the imminent possibilities of strain relief. Equally informative is the response of 1-(1-silyl)-1-(1′-2′-oxidocycloalkenyl) cyclopropanes to the action of boron trifluoride etherate (benzene, 20 °C, 2–4 hr) which produce 1-(1-silyl)-1(2-oxo cycloalkyl) cyclopropanes rather than the spirocyclononane or cyclodecane derivatives (Scheme 87b). Finally, a few reports have described the oxaspirohexane cyclopentanone transformation [218–223]. 1-(1-Seleno)cyclopropyl carbinols [87, 135] and their cyclobutyl homologs [57, 134] proved to be valuable starting materials for the ring-enlargement reactions leading to cyclobutanones and cyclopentanones, respektively. The reaction has been directly performed on 1-(1-seleno) cycloproly carbinols from which we have yet been unable to prepare oxaspiropentanes (sulfur analogs behave similarly). The formation of the cyclopentane ring from 1-(1-seleno) cyclobutylcarbinols involves the intermediate formation of oxaspirohexanes [134, 135]. These can also be prepared from the regioisomeric 1-(1-hydroxy)-1-(1′-selenoalkyl) cyclobutanes, themselves available from cyclobutanones and α-selenoalkyllithiums [44, 57, 134, 135] (Scheme 88a). As will be shown 1-(1-hydroxy)-1-(1′-selenoalkyl) cyclobutanes can also lead directly to cyclopentanones by transforming the selenenyl moiety to a better leaving group [55, 134, 135, 191, 224] (Scheme 88b).

Scheme 88

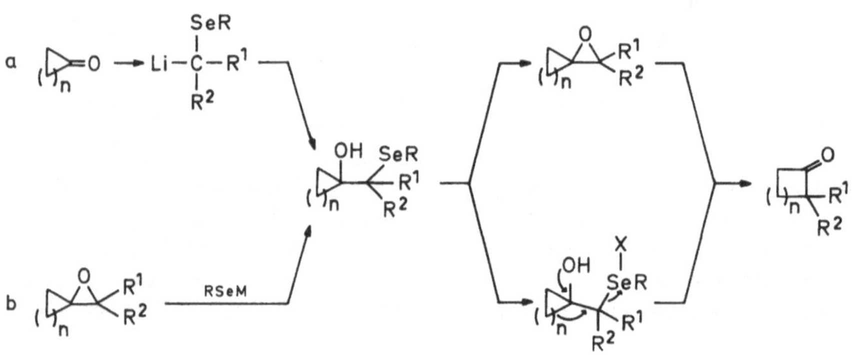

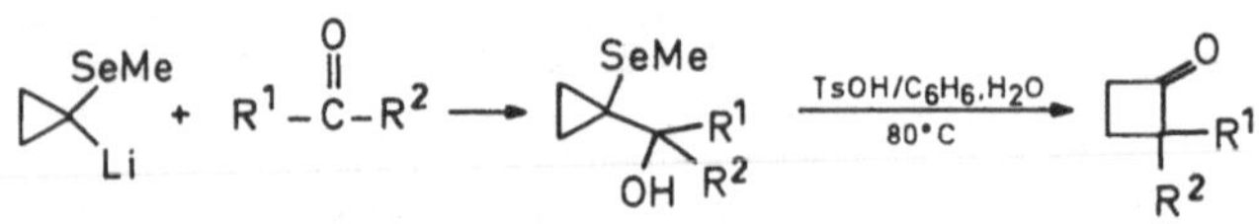

R_1	R_2	yield (%)
H	pentyl	73
H	heptyl	79
H	decyl	73
H	cyclohexyl	81
H	Ph	60
H	1-pentenyl	72
Me	H	73
Me	Ph	80
Et	Et	70
Ph	Ph	50

Scheme 89

4.5.1 Syntheses of Cyclobutanones

4.5.1.1 From β-Cyclopropylselenides and an Acid

The synthesis of cyclobutanones from 1-(1-seleno) cyclopropyl carbinols [35,87,135] parallels (Schemes 89–93) the reaction described by Trost for thiophenyl analogs [131,132]. In fact, in one instance during the synthesis of cuparenone [135] 1-(1-thiophenyl),

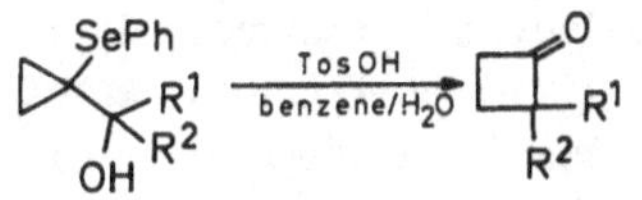

	R^1	R^2	yield
a	Me	pMePh	70%
b	H	Dec	0%
c	Me	non	0%

Scheme 90

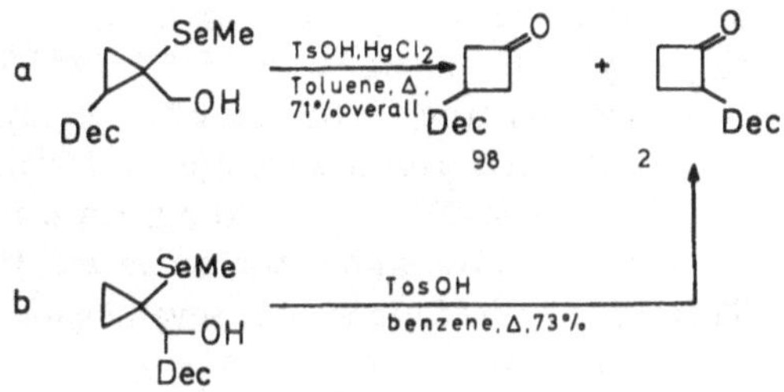

Scheme 91

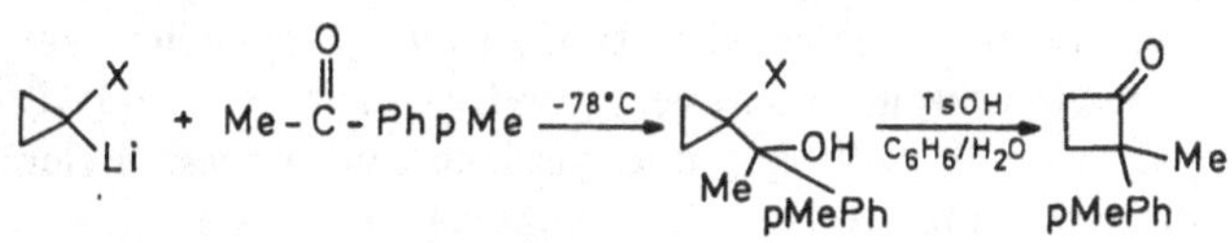

	X	solvent	yield	conditions	yield
a	SeMe	ether	85%	80°C,12h	80%
b	SePh	ether	81%	80°C,12h	70%
c	S Ph	THF	88%	40°C,12h	50%

Scheme 92

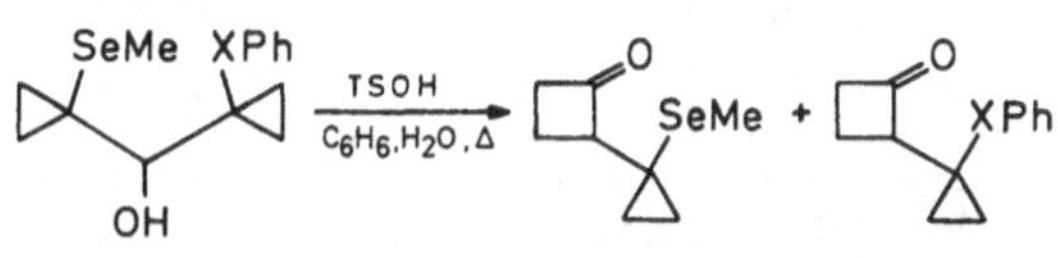

	X	overall yield	ratio	
a	S	92%	73	27
b	Se	79%	48	52

Scheme 93

1-(1-selenophenyl) and 1-(1-selenomethyl)cyclopropyl carbinols possessing the same hydrocarbon skeleton have been compared [135] and original and interesting observations were made (Scheme 92). The best results were observed [135] when the cyclopropylcarbinol belonging to the methylseleno series was heated at 80 °C in wet benzene in the presence of catalytic amounts of p-toluenesulfonic acid (Scheme 92a). It must be recalled that under these conditions olefins are usually formed [163] instead of ketones if the cyclopropane ring is missing in (β-hydroxyalkyl) selenides. The cyclobutanone synthesis accomodates [35, 87, 135] several structural variations since the rearrangement is observed with 1(1-seleno)cyclopropylcarbinols derived from aldehydes and ketones [87], including those derived from formaldehyde [35] (Scheme 91), α,β-unsaturated [35] (Scheme 89), 1-(1-seleno)cyclopropyl [35] (Scheme 93), and aromatic [135] carbonyl compounds (Scheme 92). In one case in which the cyclopropane ring was substituted with an alkyl group [35], one of the two possible regioisomeric cyclobutanones was predominantly obtained (Scheme 91). Its formation demonstrates the selective migration of the most substituted carbon present in the cyclopropane ring.

Methylseleno derivatives are clearly less reactive [135] than their phenylthio analogs, since the reactions usually require longer times and higher temperatures, but they are more than their phenylseleno analogs (Scheme 90, 92), which lead to cyclobutanones only when an extra stabilizing group such as an aryl [135] (compare Schemes 90c and 92b) or a cyclopropyl [35] (compare Schemes 90b and 93b) moiety is present on the carbon bearing the hydroxyl group. The case of the cyclopropyl compounds derived from p-methyl acetophenone, which were used as starting materials for the synthesis of cuparenone [35] (Scheme 92), is interesting, since it shows that the thiophenyl derivative is the more reactive within the series but leads [35] to a lower yield in cyclobutanone. This is due to the formation of several byproducts which were not formed from the seleno derivatives.

Dicyclopropylcarbinols bearing two different heteroatomic moieties on the cyclopropyl groups have also been subjected [35] to the ring-enlargement reaction and have lead to a mixture of cyclobutanones in very good yields (Scheme 93). It is surprising that dicyclopropylcarbinols bearing a phenylseleno and a methylseleno moiety on each of the cyclopropyl group, in contrast to what was said above, produce a 1 : 1 mixture of 2-(1-phenylseleno)cyclopropyl cyclobutanone and its methylseleno analogue (Scheme 93b). The results concerning analogous compounds bearing the methylseleno and phenylthio moieties in β- and β'-positions are closer to the predictions, since the

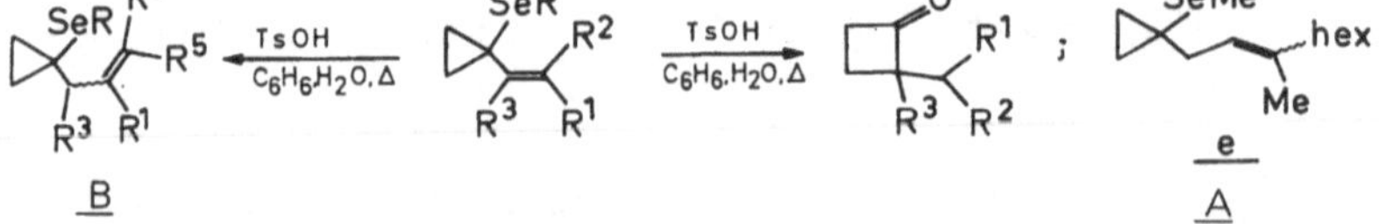

	R	R_1	R_2	R_3	yield (%) mA
a	Ph	hex	H	H	63
b	Me	hex	H	H	80
c	Me	pent	H	hex	92
d	Ph	H	H	hex	90
e	Me	Me	non	H	05*

Scheme 94

* compound B R = Me, R_1 = H, R_3 = R_4 = Me, R_5 = hex is formed along with some starting material

cyclobutanone possessing the methylseleno group clearly prevails (Scheme 93 a). Much work must be done in order to be able to draw definite conclusions on the ability of the sulfenyl and the selenenyl moieties to stabilize [15,00] an α-carbenium ion. (1-(1-seleno)-1-(1′-hydroxy-2′-seleno alkyl) cyclopropanes also produce [35] cyclobutanones on reaction with p-toluenesulfonic acid in wet benzene (Scheme 75). Surprisingly however, the selenenyl moiety expected to be there on the basis of the results reported above is now missing. These results can be explained [35,163] by the intermediate formation of a vinylcyclopropane bearing a selenenyl moiety which is in turn protonated to produce the observed cyclobutanone (Scheme 75). 1-Heterosubstituted-1-vinylcyclopropanes should therefore be valuable precursors of cyclobutanones.

4.5.1.2 From 1-Seleno-1-vinylcyclopropanes and an Acid

That is in fact the case [35,36] (Scheme 94), not only for the 1-seleno-1-vinylcyclopropanes belonging to the methylseleno series but also for their phenylseleno analogs, which both lead to cyclobutanones in the presence of p-toluenesulfonic acid on the condition [35,36] that the sp^2-carbon linked to the cyclopropane is identically or more highly substituted than the other sp^2-carbon. If that is not the case, migration of the carbon-carbon double bond from the vinylic to the allylic position (relative to the cyclopropane ring) mainly takes place and only a trace amount of cyclobutanone (<5%) and presumably cyclopentanone is formed (Scheme 94e).

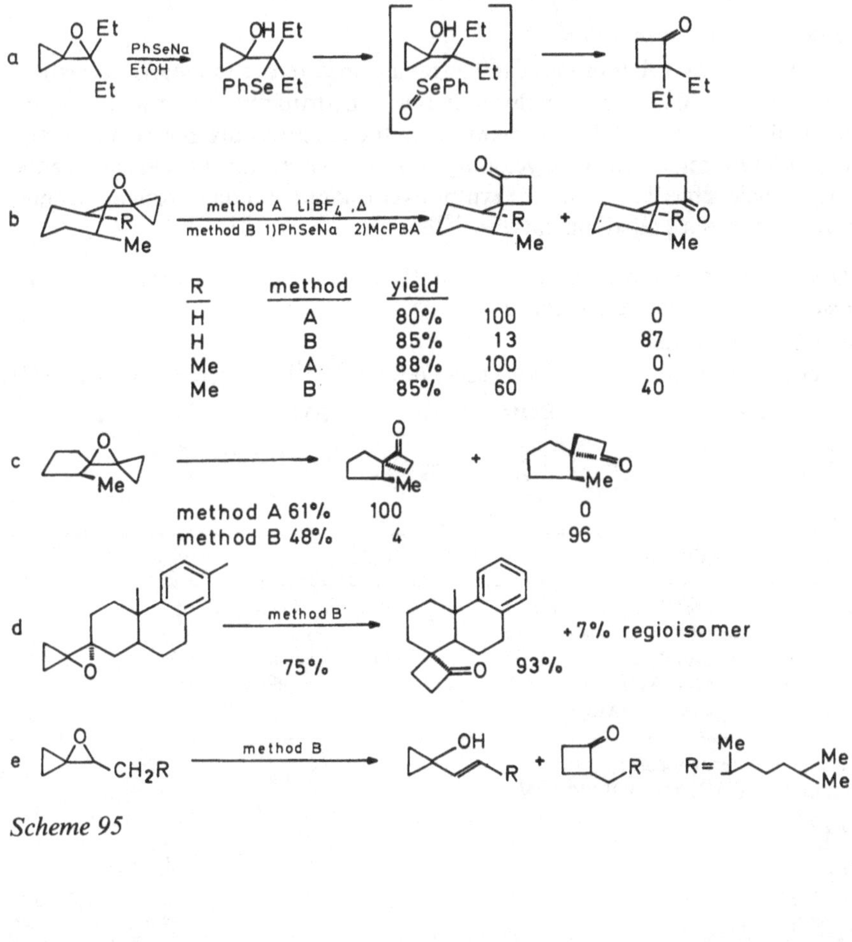

Scheme 95

4.5.1.3 From 1-(1-Hydroxy)-1-(1'-selenoalkyl) Cyclopropanes

Cyclobutanones have also been prepared[225] from 1-(1-hydroxy)-1-(1-selenoalkyl-cyclopropanes, themselves available by the ring opening of oxaspiropentanes with sodium phenylselenolate in ethanol (Scheme 95). Oxidation with m-chloroperbenzoic acid at —78 °C to —30 °C of the selenides intermediately produces the corresponding selenoxides, which in the presence of pyridine lead to cyclobutanones (Scheme 95a–d method B) by a ring-expansion reaction rather than to the allylic cyclopropanol as might be expected on the basis of previous reports on (β-hydroxyalkyl)selenoxides lacking the cyclopropane ring [3–9, 11, 12]. In fact, a mixture of allylic alchols arising from the well-known selenoxide elimination reaction and of the rearranged cyclobutanones is observed [225] with those starting materials possessing a hydrogen on the carbon bearing the seleninyl moiety (Scheme 95e), but increasing the substitution there favors the cyclobutanone formation over the allyl alcohol. A mechanism involving a carbenium ion generated by the departure of benzeneselenate has been proposed [225]. The stereochemistry of the spiro cyclobutanone seems to be determined by the rate of the ring expansion compared with that of bond rotation. Indeed, in several cases a high stereoselectivity is observed, (Scheme 95b–d method B) which is opposite to that normally produced by the acid-catalysed rearrangement of oxaspiropentanes [225] (Scheme 95b–d method A). Thus, from one oxaspiropentane either stereoisomer of cyclobutanone may be produced (Scheme 95).

4.5.2 Syntheses of Cyclopentanones

It is well known that cyclobutanones can be transformed to cyclopentane derivatives directly with diazomethane or indirectly via the cyanohydrins [218, 228], or with tris(phenylthio)methyllithium [219, 220], but in many cases the reactions are not regioselective and lead to different regioisomers. Cyclobutyl ketones lead to substituted cyclopentanones in highly acidic media [226] and the synthesis of cyclopentanones from methylenecyclobutanes requires a palladium catalyst [227].

4.5.2.1 From Oxaspirohexanes Derived from β-Hydroxy cyclobutylselenides and from β-Seleno cyclobutanols

Synthese of Oxaspirohexanes

Two different types of (β-hydroxyalkyl) selenides [35, 57, 134] or related derivatives [55],

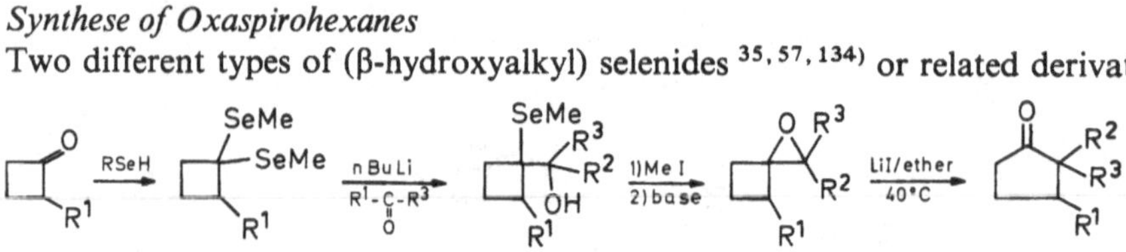

R_1	R_2	R_3	conditions	yield	time	yield
H	H	Dec	1) MeI 2) aq KOH/ether or tBuOK/DMSO	80	4 h	91[a]
H	Me	Non	1) MeI 2) aq KOH/ether	70	40 h	77[a]
Pent	H	Dec	1) MeI, $AgBF_4$ 2) aq KOH/ether or tBuOK/DMSO	71	60 h	90[b]

[a] only one stereoisomer is formed
[b] two stereoisomers (75/25 ratio) are formed

Scheme 96

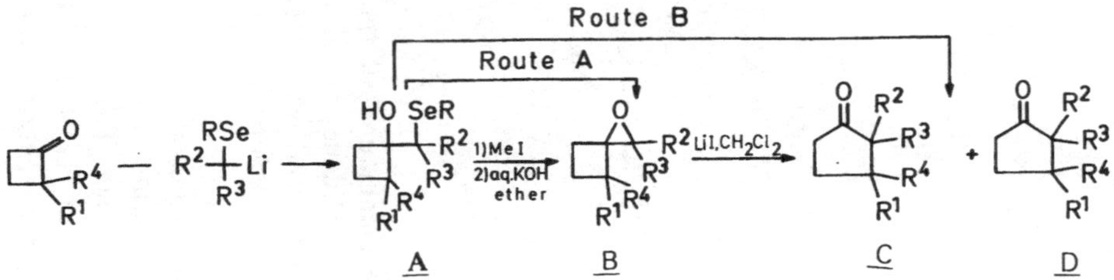

	R_1	R_4	R	R_2	R_3	solvent	A %	conditions	yield B %		C	D
a	H	H	Me	H	Dec	THF	71	Route A	65		91	—
b	Me	Non	Ph	H	H	ether	67	Route A	88	LiI, CH_2Cl_2, 40°, 60 h	83	—
c	H	H	Me	Me	Pent	THF		Route B: TlOEt/$CHCl_3$	60			—
d	H	Ph	Me	H	H	ether		Route B: MeI/CH_2Cl_2	80			—

Scheme 97

which both involve cyclobutanones as starting materials, have been prepared (Schemes 96–100). In the first case[134], the selenenyl moiety is attached to the cyclobutyl ring (Scheme 96). However, the strain present in the cyclobutane ring is not sufficient to permit (once the (β-hydroxyalkyl) selenide has reacted with para-toluenesulfonic acid in wet benzene) the ring-expansion[35] expected to lead to cyclopentanones in a reaction which would parallel the one described with

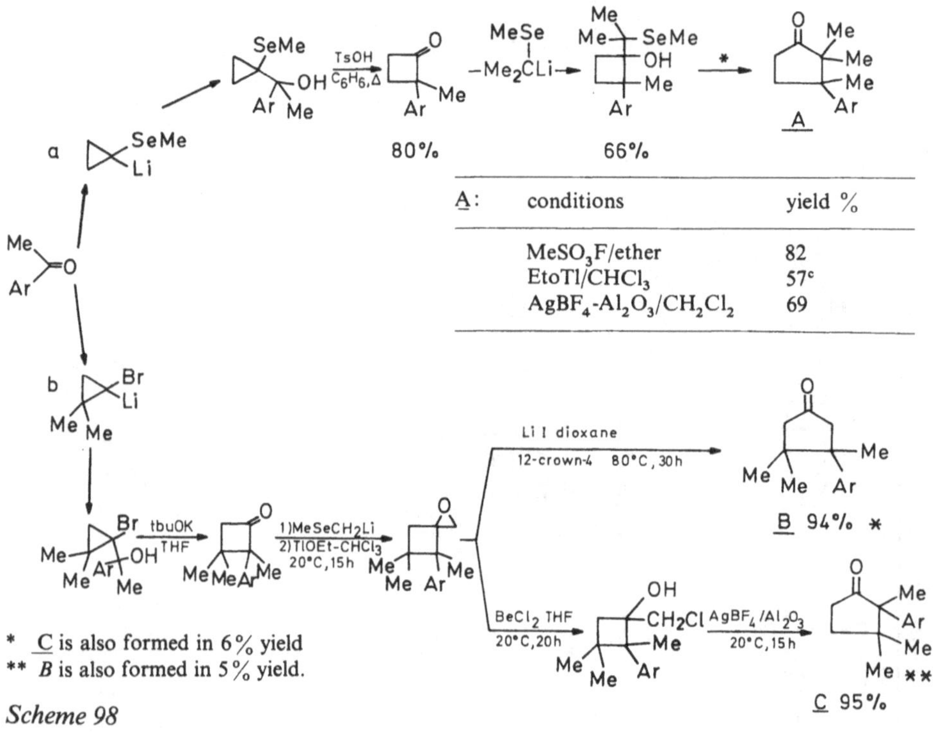

A:	conditions	yield %
	$MeSO_3F$/ether	82
	EtoTl/$CHCl_3$	57[c]
	$AgBF_4$-Al_2O_3/CH_2Cl_2	69

* C is also formed in 6% yield
** B is also formed in 5% yield.

Scheme 98

cyclopropyl analogs[35,87]. In the cyclobutyl case, in fact, the elimination of the hydroxyl and the selenenyl moieties occurs and produces alkylidene cyclobutanes[35]. The last reaction has been observed[163], as already pointed out, with other (β-hydroxyalkyl) selenides, except those derived from 1-lithio-1-cyclopropyl selenides[35,87].

The synthesis of cyclopentanones can however be readily achieved via the oxaspirohexanes[134] (Scheme 96). These are readily available from 1-(1-seleno)-1-(1′-hydroxyalkyl)cyclobutane by a route already disclosed for other (β-hydroxyalkyl) selenides[4–9,11,12,229–232] [but which does not work with 1-(1-seleno)-1-(1-hydroxyalkyl)cyclopropane]. This involves their alkylation, on the selenium atom, leading to (β-hydroxyalkyl)selenonium salts which are then almost quantitatively [134] transformed to the oxaspirohexanes on further treatment of their etheral suspension with a 10% aqueous solution of potassium hydroxide [232]. The ring-enlargement is then achieved [63, 134, 223] with lithium iodide, as will be discussed later in this section.

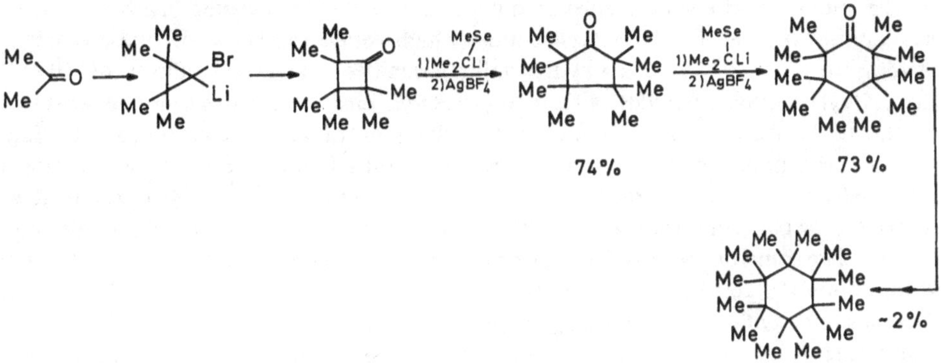

Scheme 99

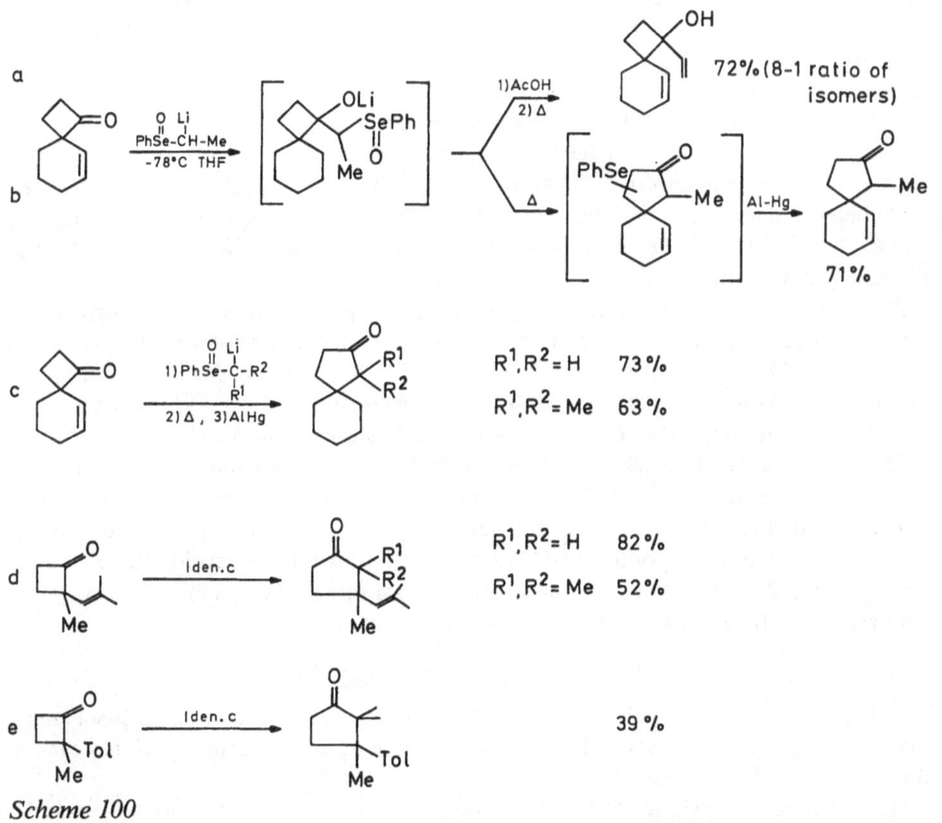

Scheme 100

In the second case, the (β-hydroxyalkyl)selenides are prepared by reaction of α-selenoalkyllithiums with cyclobutanones [35, 57, 134, 224]. Now the selenenyl moiety is no longer attached to the cyclobutyl ring (Schemes 97–99). The synthesis of oxaspirohexanes and therefore of cyclopentanones is feasible using the set of reactions already mentioned (Schemes 97, 98b) [4–9, 11, 12, 134, 229–32]. The synthesis of epoxides utilized along this novel route to cyclopentanones offers, especially in the methylseleno series, several advantages over the existing methods: wide availability of the β-selenoalkyllithiums; high nucleophilicity of these species leading to high yield of (β-hydroxyalkyl)selenides; facile alkylation of (β-hydroxyalkyl) methyl selenides with methyl iodide; easy removal by simple washing of the (β-hydroxyalkyl)selenonium salt with ether of any traces of the starting carbonyl compound which has not reacted or results from an enolisation reaction; and finally, the epoxide formed by mild treatment with potassium hydroxyde in ether is readily separated from the highly volatile dimethyl selenide concomitantly formed. The transformation of oxaspirohexanes to cyclopentanones has been achieved with lithium iodide (Schemes 97, 98b).

Transformation of Oxaspirohexanes to Cyclopentanones

The transformation of oxaspirohexanes to cyclopentanones can be achieved with lithium iodide in refluxing dichloromethane [134, 222] (Schemes 96, 97a, b). This isomerisation, first described by Lerivérend [222] and used by Trost [223] in a modified version, was previously reported for oxaspirohexanes bearing two hydrogens on the epoxide ring. It was later found in our laboratory [35, 134, 135] that it even works when two alkyl groups are attached to the epoxide ring. The reaction is usually highly regioselective and occurs by migration of the more highly alkyl-substituted carbon atom of the cyclobutane ring (Schemes 96, 97).

In the case shown in Scheme 98b, the reaction performed with LiI in CH_2Cl_2 was less regioselective [135] and although 80% of the cyclopentanone formed (B, Scheme 98) results from the migration of the most highly substituted carbon, appreciable amounts (20%) of the other regioisomer (C, Scheme 98), resulting from the migration of a methylene group, are also present.

The Trost version (LiI, benzene, eq. HMPA) [223] gave closely related results [135]. However much higher regioselectivity in favor of the β-cuparenone formation is observed if the ring-enlargement is performed with lithium iodide in dioxane, especially if a crown ether (12-Crown-4) is present in the medium [135] (Scheme 98b). Under these conditions the reaction is particularly slow (30 hrs).

The regioselective synthesis of the unnatural cuparenone isomer (C, Scheme 98) has also been performed [135] (Scheme 98b) from the same oxaspirohexane by a two-steps sequence which involves the selective (100%) — opening of the epoxide ring leading to the corresponding 1-(1-hydroxy)-1-(chloromethyl)cyclobutane ($BeCl_2$/THF, 20 °C, 20 hr) and its further transformation to the cyclopentanone C is achieved [135] with silver tetrafluoroborate on alumina.

4.5.2.2 Directly from 1-(1-Hydroxy)-1-(1′-selenoalkyl) Cyclobutanes

The direct synthesis of cyclopentanones from cyclobutanols bearing a β-selenenyl group is also possible whenever the selenenyl group is linked to a fully alkylated carbon atom (Schemes 97c, d, 98a).

This one-step procedure [43, 44, 232] is reminiscent of the well-known pinacolic rearrangement [233]. It takes advantage of the different behavior of the two heteroatomic moieties towards electrophilic species due to their hardness (hydroxyl group) or softness (selenenyl group). Thus it was found that the selenenyl group is selectively transformed, even in the presence of an hydroxyl group, to a better leaving group by alkylation, oxidation or complexation with soft metallic cations. The rearrangement of selenonium salts derived from 1-(1-hydroxy)1-(1-selenoalkyl)-cyclobutanes to cyclopentanones is particularly easy if the starting material is crowded [135,232] and possesses two alkyl groups on the carbon bearing the selenium atom (Schemes 97d, 98a). If this is not the case, the best methods are those [35,134,135] (Schemes 97c, 98a) already used successfully with other (β-hydroxylalkyl)selenides bearing two alkyl groups where the selenenyl moiety is attached, and which involve their reaction with silver tetrafluoroborate on alumina or with thallium ethoxide [44] in chloroform. In the latter case a complexed dichlorocarbene (soft reagent) is formed *in situ* and selectively acts on the selenium atom leading to an ylide. This entity acts as a base towards the hydroxyl group and at the same time enhances the leaving-group ability of the selenenyl moiety [44] (Scheme 101).

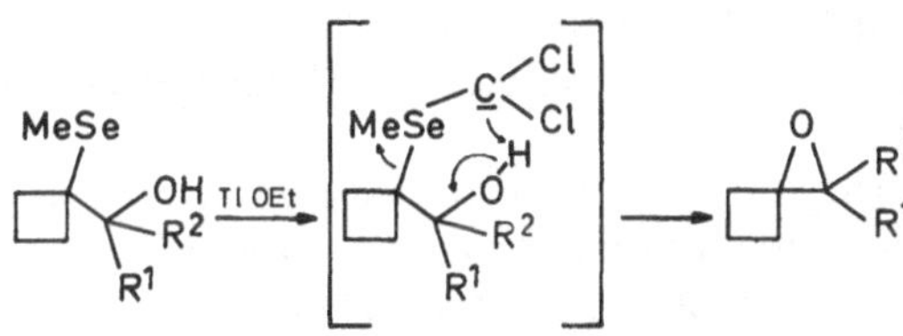

R^1	R^2	yield
H	Dec	83%
Me	Non	90%

Scheme 101

These reactions have been used for the synthesis of cuparenones from p-methoxyacetophenone [135] (Scheme 98) and for the synthesis of permethylcyclopentanone, permethylcyclohexanone, and of permethylcyclohexane from acetone [224] (Scheme 99). If the carbon bearing the selenenyl moiety bears at least one hydrogen, reaction of thallium ethoxide in chloroform is described to lead instead to epoxides [42], and this was found to be the case [234] for β-selenocylobutanols (Scheme 101). Closely related results have been observed when the reactions are performed under phase transfer catalysis simply with chloroform and potassium hydroxide as the dihalocarbene sources [235].

4.5.2.3 From 1-(1-Hydroxy)-1-(1′-selenoxyalkyl) Cyclobutanes

The reaction of α-lithioalkylselenoxides, at −78° with 2,2-disubstituted cyclobutanones affords [55] the corresponding β-alkoxy selenoxides, which, as expected [55], produce 1-(1-hydroxy)-1-(1′-alkenyl) cyclobutanes after sequential acid hydrolysis and thermolysis (Scheme 100a). However, if the acid treatment is omitted and the alkoxide is directly heated in refluxing THF, a ring-expansion takes place and leads to α-phenylselenenylated cyclopentanones by the exclusive migration of the more-substituted carbon atom (Scheme 100b–e). The formation of an α-selenenylated cyclopentanone rather than of cyclopentanone itself may account from the reaction of the last compound with a seleninyl species formed concomitantly. Cyclopentanones have been finally produced by treatment of the phenylselenenylated ketones with aluminium amalgam (Scheme 100b–e). The synthesis of various substituted cyclopentanones (Scheme 100) including cuparenone (Scheme 100e) from cyclobutanones has thus been achieved in good yields by this two-step sequence.

4.5.2.4 *Conclusion*

It is interesting to point out that the combination of cyclopropane and cyclobutane chemistry has allowed [135] the synthesis of all three regioisomers of cuparenone possessing a cyclopentanone bearing two quarternary centers in the vicinal position from p-methylacetophenone and α-selenoalkyllithiums (Scheme 98). It can be noted than α-Lithiocyclopropyl sulfides or α-lithiocyclopropyl selenides are particularly suitable when a nonalkylated cyclopropane ring is needed [135], whereas α-lithio-α-bromo cyclopropanes are the reagents of choice [135, 224] for the introduction of the polyalkylated cyclopropane moiety.

The very high nucleophilicity of α-heterosubstituted α-lithiocyclopropanes and of α-lithioalkyl selenides, even those bearing two alkyl groups on the carbanionic center, permits the stepwise construction of a cyclopentane ring possessing several quaternary centers in vicinal positions with respect to one another by two successive ring expansion reactions from a suitably functionalyzed cyclopropane [135, 224] (Schemes 98, 101). This feature has been used by Fitjer [224] for the synthesis of permethyl cyclobutanone, permethyl cyclopentanone, and even permethyl cyclohexanone, as well as for the preparation of previously unknown permethylcyclohexane.

5 Summary

α-Selenocyclopropyllithiums, α-selenocyclobutyllithium, and α-silylcyclopropyllithiums are versatile and easily prepared intermediates which allow a large variety of reactions. Some permit the formation of functionalized derivatives which still posses the strained ring present in the starting materials, such as alkylidene cyclocyclopropanes and alkylidene cyclobutanes, cyclobutenes, functionalized vinylcyclopropanes and vinylcyclobutanes as well as oxaspirohexanes. The strain present in the small ring systems can also be released. This has been used, inter alia, for the synthesis of cyclobutanones from cyclopropyl derivatives and cyclopentanones from cyclobutyl derivatives. The process can be repeated in some instances to homologate cycloalkanones. Cyclobutyl derivatives also allow the synthesis of dienes, inculding 2-functionalized ones. Allylidene cyclopropanes have been found powerful partners in Diels-Alder reactions.

All these compounds can themselves be used in further transformations.

6 References

1. Rheinboldt H., Houben-Weyl: In Methoden der Organischen Chemie, Müller E. G. (eds.) Thieme Verlag, Stuttgart, *IX*, 917 1967
2. Klayman D. L., Gunther W. H. H.: Organic Selenium Compounds: Their Chemistry and Biology, Klayman D. L., Günther W. H. H. (eds.), Wiley, New York 1973
3. Sharpless K. B., Gordon K. M., Lauer R. F., Singer S. P., Young M. W.: Chem. Scr. *8 A*, 9 (1975)
4. Clive D. L. J.: Tetrahedron *34*, 1049 (1978)
5. Reich H. J.: Acc. Chem. Res. *12*, 22 (1979)
6. Reich H. J.: Oxidation in Organic Chemistry. in "Organic Chemistry a Series of monographs". Trahanovsky W. S. (ed.) Academic Press, New York vol. *5C*, 1 (1978)
7. Krief A.: Tetrahedron *36*, 2531 (1980)
8. Krief A, In Organic Chemistry of Selenium and Tellurium Containing Fonctional Groups., Patai S., Rappoport Z. (eds.), Wiley, London in the Press
9. Krief A., Hevesi L.: Applications of Selenium chemistry to Organic Synthesis, Springer-Verlag, Berlin, Heidelberg in the Press
10. Hevesi L.: In Organic Chemistry of Selenium and Tellurium Containing Fonctional Groups, Patai S., Rappoport Z. (eds.). Wiley. London 1986
11. Nicolaou K. C., Petasis N. A.: Selenium in Natural Products Synthesis, CIS, Inc., Philadelphia 1984
12. Krief A., Hevesi L.: Janssen Chimica Acta *2*, 3 (1984)
13. Krief A., Dumont W., Denis J. N., Evrard G., Norberg B.: J. Chem. Soc., Chem. Commun. 569 (1985)
14. Krief A., Dumont W., Denis J. N.: J. Chem. Soc., Chem. Commun. 571 (1985)
15. Nsunda K. M., Hevesi L.: Tetrahedron Lett. *25*, 4441 (1984)
16. Sevrin M., Van Ende D., Krief A.: Tetrahedron Lett. 2643 (1976)
17. Clive D. L. J., Chittattu G., Wong C. K.: J. Chem. Soc., Chem. Commun. 41 (1978)
18. Clive D. L. J., Chittattu G. J., Farina V., Kiel W. A., Menchen S. M., Russel C. G., Singh A., Wong C. K., Curtis N. J.: J. Am. Chem. Soc. *102*, 4438 (1980)
19. Sevrin M., Dumont W., Hevesi L., Krief A.: Tetrahedron Lett. 2647 (1976)
20. Sevrin M., Krief A.: J. Chem. Soc., Chem. Commun. 656 (1980)
21. Hevesi L., Sevrin M., Krief A.: Tetrahedron Lett. 2651 (1976)
22. Kingsbury C. A., Cram D. J.: J. Am. Chem. Soc. *82*, 1810 (1960)
23. Halazy S., Krief A.: Tetrahedron Lett. 4233 (1979)
24. Sharpless K. B., Lauer R. F.: J. Org. Chem. *37*, 3973 (1972)
25. Di Giamberardino T., Halazy S., Dumont W., Krief A.: Tetrahedron Lett. *24*, 3413 (1983)
26. Sharpless K. B., Lauer R. F., J. Amer. Chem. Soc. *94*, 7154 (1972)
27. Wilson C. A., Bryson T. A.: J. Org. Chem. *40*, 800 (1975)

28a Clive D. L. J., Chittattu G., Curtis N. J., Menchen S. M.: J. Chem. Soc. 770 (1978)

28b Clive D. L. J. Tetrahedron, *34*, 1049 (1978)

29. Reich H. J.: J. Org. Chem. *40*, 2570 (1975)
30. Salmond W. G., Barta M. A., Cain A. M., Sobala M. C.: Tetrahedron Lett. 1683 (1977)
31. Halazy S., Krief A.: Tetrahedron Lett. *22*, 1833 (1981)
32. Nyshiyama H., Itagaki K., Sakuta K., Itoh K.: Tetrahedron Lett. *22*, 3285 (1981)
33. Nyshiyama H., Itagaki K., Osaka N., Itoh K.: Tetrahedron Lett. *23*, 4103 (1982)
34. Nishiyama K., Kitajima T., Yamamoto A., Itoh K.: J. Chem. Soc., Chem. Commun. 1232 1982)
35. Halazy S.: PhD. Thesis, Faculté N. D. de la Paix, Namur (1982)
36. Halazy S., Krief A.: Tetrahedron Lett. *22*, 4341 (1981)
37. Halazy S., Krief A.: Tetrahedron Lett. *22*, 1829 (1981)
38. Fankhauser J. E., Peevey R. M., Hopkins P. B.: Tetrahedron Lett. *25*, 15 (1984)
39. Reich H. J., Chow F., Shah S. K.: J. Am. Chem. Soc. *101*, 6638 (1979)
40. Clarembeau M., Krief A.: Tetrahedron Lett. *25*, 3629 (1984)
41. Sharpless K. B., Lauer R. F.: J. Am. Chem. Soc. *94*, 7154 (1972)
42. Laboureur J. L., Dumont W., Krief A.: Tetrahedron Lett. *25*, 4569 (1984)
43. Labar D., Laboureur J. L., Krief A.: Tetrahedron Lett. *23*, 983 (1982)
44. Laboureur J. L., Krief A.: Tetrahedron Lett. *25*, 2713 (1984)

45. Halazy S., Krief A.: Tetrahedron Lett. *22*, 2135 (1981)
46. Seebach D., Peleties N.: Angew. Chem., Int. Ed. Engl. *8*, 450 (1969)
47. Seebach D., Peleties N.: Chem. Ber. *105*, 511 (1972)
48. Van Ende D., Cravador A., Krief A.: J. Organomet. Chem. *177*, 1 (1979)
49. Raucher S., Koolpe G. A.: J. Org. Chem. *43*, 3794 (1978)
50. Reich H. J., Willis W. W., Clark P. D.: J. Org. Chem. *46*, 2775 (1981)
51. Gröbel B. T., Seebach D.: Chem. Ber. *110*, 852 (1977)
52. Van Ende D., Dumont W., Krief A.: J. Organoment. Chem.,*C10*, 149 (1978)
53. Reich H. J., Shah S. K.: J. Am. Chem. Soc. *97*, 3250 (1975)
54. Reich H. J., Shah S. K., Chow F.: J. Am. Chem. Soc., *101*, 6648 (1979)
55. Gadwood R. C.: J. Org. Chem. *48*, 2098 (1983)
56. Clarembeau M., Cravador A., Dumont W., Hevesi L., Krief A. Lucchetti J., Van Ende D.: Tetrahedron *41*, 4793 (1985)
57. Halazy S., Krief A.: Tetrahedron Lett. *21*, 1997 (1980)
58. Sisido K., Utimoto K.: Tetrahedron Lett. 3267 (1966)
59. Bestman H. J., Denzel T.: Tetrahedron Lett. 3591 (1966)
60. Schweizer E. E., Berninger C. J., Thompson J. G.: J. Org. Chem., *33*, 336 (1968)
61a Trost B. M., Bogdanowicz M. J.: J. Am. Chem. Soc. *93*, 3773 (1971)
61b Trost B. M., Bogdanowicz M. J.: J. Am. Chem. Soc. *95*, 5311 (1973)
61c Trost B. M., Bogdanowicz M. J.: J. Am. Chem. Soc. *95*, 5298 (1973)
62. Trost B. M., Melvin L. S.: Sulfur Ylides Emerging Synthetic Intermediates in "Organic Chemistry a Series of monographs". A. T. Blomquist, Wassermann H. H. (eds.) Academic Press. N. Y. *31* (1975)
63. Johnson C. R., Katekar G. F., Huxol R. F., Janiga E. R.: J. Am. Chem. Soc. *93*, 3771 (1971)
64. Wiseman J. R., Chan H. F.: J. Amer. Chem Soc *92*, 4749/1970)
65. Truce W. E., Lindy L. B.: J. Org. Chem. *26*, 1463 (1961)
66. Chang Y., Pinnick H. W.: J. Org. Chem., *43*, 373 (1978)
67. Dumont W., Krief A.: unpublished results (1985)
68. Walborsky H. M., Perrasany M. P.: J. Am. Chem. Soc. *96*, 3711 (1974)
69. Harms R., Schöllkopf U., Muramatsu M.: Liebigs Ann. Chem. 1194 (1974)
70. Trost B. M., Keeley D. E., Arndt H. C., Rigby J. H., Bogdanowicz M. J.: J. Am. Chem. Soc. *99*, 3080 (1977)
71. Braun M., Damman R., Seebach D.: Chem. Ber. *108*, 2368 (1975)
72. Lambert R. L., Seyferth D.: J. Am. Chem. Soc. *94*, 9246 (1972)
73. Hiyama T., Takehara S., Kitatani K., Nozaki H.: Tetrahedron Lett. 3295 (1974)
74. Kobrich G., Goyert W.: Tetrahedron, 4327 (1968)
75. Seyferth D., Duncan P.: J. Organomet. Chem. *111*, C21 (1976)
76. Cunico R. F., Han Y. K.: J. Organomet. Chem. *174*, 247 (1979)
77. Halazy S., Dumont W., Krief A.: Tetrahedron Lett. *22*, 4737 (1981)
78. Hiyama H., Kanakura A., Morizawa Y., Nozaki H.: Tetrahedron Lett. *23*, 1279 (1982)
79. Cohen T., Daniewski W. M., Weisenfeld R. B.: Tetrahedron Lett. 4665 (1978)
80. Cohen T., Matz J. R.: Synth. Commun. *10*, 311 (1980)
81a Cohen T., Sherbine J. P., Matz J. R., Hutchins R. R., McHenry B. M., Willey P. R.: J. Am. Chem. Soc. *106*, 3245 (1984)
81b. Cohen T., Sherbine J. P., Mendelson S. A., Myers M. Tetrahedron Lett. *26*, 2965/1985
82. Paquette L. A., Horn K. A., Wells G. J.: Tetrahedron Lett. *23*, 259 (1982)
83. Seyferth D., Lambert R. L.: J. Organomet. Chem. *88*, 287 (1975)
84. Cohen T., Matz J. R.: J. Am. Chem. Soc. *102*, 6900 (1980)
85. Cohen T., Matz J. R.: Tetrahedron Lett. *22*, 2455 (1981)
86. Halazy S., Lucchetti J., Krief A.: Tetrahedron Lett. 3971 (1978)
87. Halazy S., Krief A.: J. Chem. Soc., Chem. Commun. 1136 (1979)
88. Cram D. J.: Fundamentals of Carbanion Chemistry in "Organic Chemistry a Series of monographs". Academic Press *4*, 49 (1965)
89. Rol N. C., Clague D. H.: Org. Magn. Reson. *16*, 187 (1981)
90. Müller N., Putchard D.: J. Chem. Phys. *31*, 768 (1959)
91. Patel D. J., Howden M. E. H. Roberts J. D.: J. Am. Chem. Soc. *85*, 3218 (1963)
92. Shono T., Morikawa T. Oku A., Oda R.: Tetrahedron Lett. 791 (1964)

93. Charton M., Zabricky. The chemistry of Alkenes. Intersciences *2*, 524 (1970)
94. Halazy S., Dumont W., Krief A.: unpublished results (1985)
95. Schöllkopf U., Küppers H.: Tetrahedron Lett. 105 (1963)
96. Seyferth D., Cohen H. M.: Inorg. Chem. *1*, 913 (1962)
97. Sakurai H., Hosomi A., Kumada M.: Tetrahedron Lett. 2469 (1968)
98. Cunico R. F.: cited in our ref. 82.
99. Clarembeau M., Krief A.: unpublished results (1985)
100. Schmidt A., Kobrich G.: Tetrahedron Lett. 2561 (1974)
101. Akbachinskaya T. V., Bakhbulk M., Grishin Y. K., Donskaya N. A., Ustynyuk Y. A. zh. org. Khim. *14*, 2317 (1978)
102. Moss R. A., Munjal R. C.: Synthesis 425 (1979)
103. Schlosser M., Schneider P.: Helv. Chim. Acta *63*, 2404 (1980)
104. Donskaya N. A., Akhachinskaya T. V., Leonova T. V., Shulishov E. V., Shabarov Y. S.: Zh. Org. Khim. *16*, 563 (1980)
105. Dumont W., Krief A.: Angew. Chem. Int. Ed. Engl. 161 (1976)
106. Clarembeau M., Bertrand J. L., Krief A.: Isr. J. Chem. *24*, 125 (1984)
107. Zutterman F., Krief A.: Org. Chem. *48*, 1135 (1983)
108. Lucchetti J., Remion J., Krief A.: C. R. Acad. Sci., Ser. C *288*, 553 (1979)
109. Braun M., Seebach D.: Chem. Ber. *109*, 669 (1976)
110. Feugeas C., Galy J. P.: C. R. Acad. Sci., Ser. *C270*, 2157 (1970)
111. Cohen T., Daniewski W. N.: Tetrahedon Lett. 2991 (1978)
112. Rühlmann K.: Synthesis 236 (1971)
113. Jooritsma R.: Thèse de doctorat, Amsterdam 1979
114. Denis J-N.: Research Report, Namur 1983
115. Rousseau G., Sloughi N.: Tetrahedon Lett. *24* 1251 (1983)
116. Schauder J-R., Krief A.: Research Report, Namur 1984
117. Di Giamberardino T.: Mémoire d'ingénieur, Namur (1981)
118. Chan T. H.: Acc. Chem. Res., *10*, 442 (1977)
119. Huldrik P. F., Peterson D.: J. Am. Chem. Soc. *97*, 1464 (1975)
120. Chan T. H., Massuda D.: Tetrahedron Lett. 3383 (1975)
121. Chan T. H., Mychajlowskij W., Ong. B. S., Harpp D. N.: J. Org. Chem. *43*, 1526 (1978)
122. Ho T-L: J. Chem. Educ. 355 (1978)
123. Léonard-Coppens A. M., Krief A.: Tetrahedron Lett. 3227 (1966)
124. Cram D. J., Ahmed Abd Elhafez F.: J. Am. Chem. Soc. *74*, 5828 (1952)
125. Cherest M., Felkin H., Prudent N.: Tetrahedron Lett. 2199 (1968)
126. Hassig R., Siegel H., Seebach D.: Chem. Ber. *115*, 1990 (1982)
127. Kitatani K., Hiyama T., Nozaki H.: J. Am. Chem. Soc. *97*, 949 (1975)
128. Moore W. R., Ward H. R.: J. Org. Chem. *27*, 4179 (1962)
129. Skatteböl L.: Acta Scand. 1683 (1963)
130. Sandler S. R., Karo W.: In Organic Functional Group Preparations. In "Organic chemistry a series of monographs. Blomquist A. T., (ed.), Academic Press, New York, *2*
131. Trost B. M., Keeley D. E., Arndt H. C., Bogdanowicz M. J.: Am. Chem. Soc. *99*, 3088 (1977)
132. Trost B. M.: Acc. Chem. Res. *7*, 85 (1974)
133. Trost B. M.: Pure Appl. Chem. *43*, 563 (1975)
134. Halazy S., Krief A.: J. Chem. Soc., Chem. Commun. 1200 (1982)
135. Halazy S., Zutterman F., Krief A.: Tetrahedron Lett. *23*, 4385 (1982)
136. Paquette L. A., Wells G. T., Wickham G.: J. Org. Chem. *49*, 3618 (1984)
137. Paquette L. A., Wells G. J., Horn K. A., Yan T-H.: Tetrahedron Lett. *23*, 263 (1982)
138. Paquette L. A., Yan T-H., Wells G. J.: J. Org. Chem. *49*, 3610 (1984)
139. Masuyama Y., Ueno Y., Okawara M.: Chem. Lett. *7*, 835 (1977)
140. Van Den Heuvel C. J. M.: These de Doctorat, Amsterdam 1980
141. Mortillaro L., Credali L., Mammi M., Valle G.: J. Chem. Soc. 807 (1965)
142. Eyman W. Hanack M.: Tetrahedron Lett. 4213 (1972)
143. Bertrand M., Monti H.: C. R. Acad. Sci., Ser. C *264*, 998 (1967)
144. Schipperyn A. J., Smack P.: Recl. Trav. Chim. Pays-Bas *92*, 1158 (1973)
145. Vincens M., Dumont C., Vidal M.: Bull. Soc. Chim. Fr. *12*, 2811 (1974)

146. Cope A. C., Ambros D., Ciganek E., Howell C. F., Jacura Z.: J. Am. Chem. Soc. *82*, 1750 (1960)
147. Wiberg K. B., Fenoglio R. A.: J. Am. Chem. Soc. *90*, 3395 (1968)
148. Shand W., Schomaker V., Fisher J. R.: J. Am. Chem. Soc. *66*, 636 (1944)
149. Gil-Av E., Herling J.: Tetrahedron Lett. 27 (1961)
150. Shabtai J., Gil-Av E.: J. Org. Chem. *28*, 2893 (1963)
151. Sevrin M., Krief A.: Tetrahedron Lett. 187 (1978)
152. Vogel E.: Angew. Chem. *66* 640 (1954)
153. Cooper W., Walters W. D.: J. Am. Chem. Soc. *80*, 4220 (1958)
154. Frey H. M.: J. Chem. Soc., Chem. Commun. 957 (1961)
155. Woodward R. B., Hoffmann R.: The Conservation of Orbital Symmetry Academic Press. N.Y. (1970)
156. Carey R. A., Sundberg R. S.: Advanced Organic chemistry. Plenum Press. N. Y. (1977)
157. Woodward R. B., Hoffmann R.: J. Am. Chem. Soc. *87*, 395 (1965)
158. Frey H. M., Walsh R.: Chem. Rev. 103 (1969)
159. Wilson S. R., Phillips L. R.: Tetrahedron Lett. 3047 (1975)
160. Wilson S. R., Phillips L. R., Nathalie K. J.: J. Am. Chem. Soc. *101*, 3340 (1979)
161. Trost B. M., Vladuchik W. C., Bridges A. J.: J. Am. Chem. Soc. *102*, 3548 (1980)
162. Jefford C. W., Boshung A. F., Rimbault C. G.: Tetrahedron Lett 3387 (1974)
163a. Remion J., Dumont W., Krief A.: Tetrahedron Lett. 1385 (1976)
163b. Remion J., Krief A.: Tetrahedron Lett., 3743 (1976)
164. Stirling C. J. M.: Isr. J. Chem., *21*, 111 (1981)
165. Trost B. M., Keeley D. E.: J. Am. Chem. Soc. *98*, 248 (1976)
166. Julia M., Julia S., Guégan R.: Bull. Soc. Chim. Fr. 1072 (1960)
167. Brady S. F., Ilton M. A., Johnson W. S.: J. Am. Chem. Soc. *90*, 2882 (1968)
168. Denis J. N., Dumont W., Krief A.: Tetrahedron Lett. 4111 (1979)
169. Dumont W., Krief A.: J. Chem. Soc., Chem. Commun. 673 (1980)
170. Denis J. N., Krief A.: J. Chem. Soc., Chem. Commun. 229 (1983)
171. Flowers M. C., Frey H. M.: J. Chem. Soc. 3547 (1961)
172. Brulé D., Chalchat J. C., Garry R. P., Lacroix B., Michet A. and Vessière R., Bull. Soc. Chim. Fr., *2*, 57 (1981)
173. Andrews G. D. and Baldwin J. E., J. Am. Chem. Soc., *98*, 6705 (1976)
174. Ellis R. J. and Frey H. M., J. Chem. Soc., 5578 (1964).
175. Ellis R. J. and Frey H. M., J. Chem. Soc., 4188 (1964).
176. Kierstead R. W., Linstead R. P., Weedom B. C. L.: J. Chem. Soc. 3616 (1952)
177. Davrand G., Miginiac P.: Tetrahedron Lett. 997 (1972)
178. Miyaura N., Itoh M., Sasaki N., Suzuki A.: Synthesis 317 (1975)
179. Grieco P., Finkelhor R.: J. Org. chem. *38*, 2100 (1973)
180. Pereyre M., Quintard J. P.: Pure Appl. Chem. *53*, 2401 (1981)
181. Halazy S.: unpublished results (1985)
182. Wells G. J., Yan T-H., Paquette L. A.: J. Org. Chem. *49*, 3604 (1984)
183. Burgess E. M., Penton H. R., Taylor E. A.: J. Org. Chem. *38*, 26 (1973)
184. Franck-Neumann M., Lohmann J. J.: Angew. Chem. Int. Ed. Engl. *16*, 323 (1977)
185. Denis J. N., Desauvage S., Hevesi L., Krief A.: Tetrahedron Lett. *22*, 4009 (1981)
186. Paquette L. A.: The Ramberg-Bläcklund Rearrangement: Organic Reactions *25*, 1 (1977)
187. McMurry J. E., Fleming M. P., Kees K. L. J., Krepski L. R.: J. Org. Chem. *43*, 3255 (1978)
188. McMurry J. E., Krepski L. R.: J. Org. Chem. *41*, 3929 (1976)
189. Sharpless K. B., Young M. W., Lauer R. F.: Tetrahedron Lett. 1979 (1973)
190. Evans D. A., Andrews G. C.: Acc. Chem. Res. *7*, 147 (1974)
191. Neureiler N. P.: J. Org. Chem. *24*, 2044 (1959)
192. Overberger C. G., Borchert A. E.: J. Am. Chem. Soc. *82*, 4896 (1960)
193. Doering W. V. E., Sachdev K.: J. Am. Chem. Soc. *96* 1168 (1974)
194. Doering W. V. E., Sachdev K.: J. Am. Chem. Soc. *97*, 5512 (1975)
195. Piers E., Banville J.: J. Chem. Soc., Chem. Commun. 1138 (1979)
196. Wender P. A. and Filosa M. P., J. Org. Chem., *41*, 3490 (1976)
197. Hudlicky T. H., Kutchan T. M., Wilson S. R., Mao D. T.: J. Am. Chem. Soc. *102*, 6351 (1980)
198. Morton H. E.: J. Org. Chem. *43*, 3630 (1978)

199. Marino J. P., Browne M. J.: Tetrahedron Lett. 3245 (1976)
200. Mil'vitskaya E. M., Tarakanova A. V., Plate A. F.: Russ. Chem. Rev. *45*, 469 (1976)
201. Fleming I: in Comprehensive Organic Chemistry, Barton D. H. R., Ollis W. D. (eds): Pergamon Press (1979)
202. Chan T. H., Fleming I.: Synthesis 761 (1979)
203. Zutterman F.: Research Report, Namur 1982
204. Roth W. R., Schmidt T.: Tetrahedron Lett. 3639 (1971)
205. Kende A. S., Rieke E. E.: J. Am. Chem. Soc. *94*, 1397 (1972)
206. March J.: Advanced Organic Chemistry: Reactions, Mechanisms and Structure, McGraw-Hill, Tokyo, 1977, p. 761
207. Wollweber H.: Dields-Alder-Reaktion, in Methoden der Organischen Chemie Müller, E. G. Thieme Verlag, Stuttgart, 5/1C, 981 (1970)
208. Steward C. A.: J. Am. Chem. Soc. *28*, 3320 (1963)
209. Stewart C. A.: J. Am. Chem. Soc. *84*, 1117 (1962)
210. Slobodin Y. M., Grigoreva V. I., Smulyakovskii Y. E.: Zh. Obshch. Khim. *23*, 1873 (1958)
211. Goldman N. L.: Chem. Ind. (London) 1036 (1963)
212. Ichikizaki I., Avai A.: Bull. Chem. Soc. *37*, 432 (1964)
213. See ref 5 described in our ref 107,
214. Zutterman F.: Research Report, Namur 1983
215. Eisenstein O., Lefour J. M., Nguyen Trong Anh: Tetrahedron *33*, 523 (1970)
216. Sarel S., Yovell J., Sarel-Imber M.: Angew. Chem. Int. Ed. Engl. *7*, 577 (1968)
217. Wenkert E.: Acc. Chem. Res. 27 (1980)
218. Gutsche C. D., Redmore D.: Carbocylic ring Expansion Reactions Academic Press, New York 1968
219. Cohen T., Kuhn D., Falck J. R.: J. Am. Chem. Soc. *97*, 4749 (1975)
220. Knapp S., Trope A. F., Ornaf R. M.: Tetrahedron Lett *21*, 4301 (1980)
221. Ogura K., Yamashita M., Suzuki M., Tsuchiashi G.: Chem. Lett. 93 (1982)
222. Lerivérend M. L., Lerivérend P.: C. R. Acad. Sci., Ser. C *280C*, 791 (1975)
223. Trost B. M., Latime L. H.: J. Org. Chem. *43*, 1031 (1978)
224. Fitzer L., Scheuermann H-J., Wehle D.: Tetrahedron Lett *25*, 2329 (1984)
225. Trost B. M., Scudder P. H.: J. Am. Chem. Soc. *99*, 7601 (1977)
226. Larchevèque M., Mulot P., Cuvigny T.: C. R. Acad. Sci., Ser. *C 280 C*, 309 (1975)
227. Grugg R., Boontanonda P.: J. Chem. Soc., Chem. Commun. 583 (1977)
228. Gutsche C. D.: Carbocyclic Ring Expansion Reactions Academic Press New York (1968)
229. Dumont W., Krief A.: Angew. Chem. Int. Ed. Engl. *14*, 350 (1975)
230. Van Ende D., Dumont W., Krief A.: Angew. Chem. Int. Ed. Engl. *14*, 700 (1975)
231. Van Ende D., Krief A.: Tetrahedron Lett. 457 (1976)
232. Labar D., Krief A.: J. Chem. Soc., Chem. Commun. 564 (1982)
233. Suzuki K., Katayama E., Tsuchiashi G.: Tetrahedron Lett. *25*, 1817 (1984) and references cited
234. Halazy S.: Research Report (1982)
235. Laboureur J. L., Krief A.: unpublished results (1985)

Cyclopropenes and Methylenecylopropanes as Multifunctional Reagents in Transition Metal Catalyzed Reactions

Paul Binger and Holger Michael Büch*

Max-Planck-Institut für Kohlenforschung, Postfach 101353
D-4330 Mülheim a. d. Ruhr, West Germany (FRG)

Table of Contents

---

* present address:
Farbwerke Hoechst AG, D-6230 Frankfurt/Main 80

The tremendous potential of cyclopropenes and methylene-cyclopropanes as multifunctional reagents in organic syntheses has only been recognized in the last decade. The use of transition metal catalysts allows an effective control of their transformations enabling highly selecitive syntheses ranging from three- to sevenmembered carbocycles. The activities in this field are summarized in this review for the first time with an emphasis on the preparative aspects of this work.

1 Introduction

Some 15 years ago, convenient syntheses of methylenecyclopropane [1-5], organo-substituted methylenecyclopropanes [6] and organo-substituted cyclopropenes [7-10] have been developed. Most of these highly strained molecules are surprisingly stable. For example, methylenecyclopropane (b.p. 11 °C), which is now available on a kilogram scale [5], can be stored in a cylinder at room temperature for several years without significant decomposition. Therefore these molecules are not only suited for mechanistic studies but also for synthetic applications. Cyclopropenes and methylenecyclopropanes are potentially multifunctional and thus should be very versatile in organic synthesis (cf. Fig. 1).

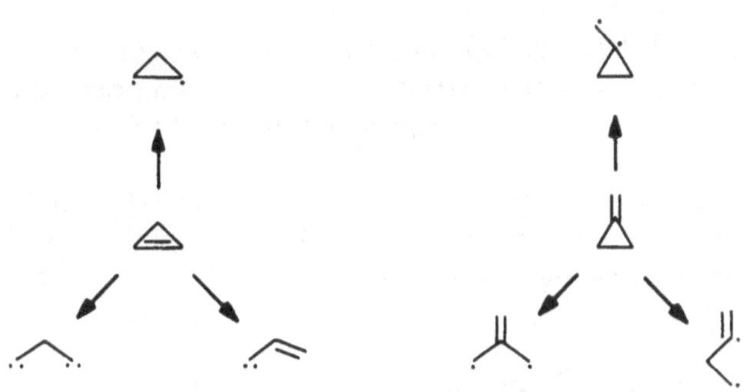

Fig. 1. Synthons derived from cyclopropenes and methylenecyclopropanes via transition metal catalysis (dipolar structures not considered)

The general tendency of three-membered heterocycles, to react via open chain (1,3-dipolar) isomers [11,12], is less pronounced for three-membered carbocycles.

Thus, the traditional organic chemistry of cyclopropenes and methylenecyclopropanes is mainly that of reactive alkenes [13-15,128]. This drawback, however, can be overcome by the use of suitable transition metal catalysts.

Moreover, modification (tailoring) of these catalysts allows a fine tuning of the selectivity of the reactions which is almost unique in organic synthesis. As an example par excellence, one may consider the cyclooligomerization of 3,3-dimethylcyclopropene. In the presence of phosphane-modified Pd(0) catalysts, the cyclopropene is cyclotrimerized quantitatively and stereoselectively, whereas with the aid of phosphane-free Pd(0) catalysts a single cyclodimer is obtained in high yield.

The synthetic potential of transition metal catalyzed transformations of unsaturated three-membered carbocycles has just begun to emerge. Selective cycloaddition methods providing three- to eight-membered rings of different functionality are available. The cooligomerization of methylenecyclopropanes with a number of heterocumulenes to give heterocycles has only recently been uncovered and is another exciting aspect of the chemistry. Ring-opening reactions can either be used to build up new carbocycles or to synthesize open-chain polyenes in one step. Although our understanding of the mechanistic details of these catalytic processes is far from being complete, it is high time for a comprehensive review of the field. In the following we will discuss the results obtained so far with an emphasis on the preparative aspects. One of the goals of this article is to propagate the use of cyclopropenes and methylenecyclopropanes as low-cost, multifunctional reagents. Since the handling of air- and moisture-sensitive compounds is common practice nowadays, the combination of transition metal catalysts with strained three-membered carbocycles opens new perspectives for selective C—C and C—X bond formations.

2 Reactions of Cyclopropenes

2.1 General Considerations

Preparation [13–15] and properties [13–17] of cyclopropenes have been reviewed extensively together with different aspects of their chemistry [13–15, 18–23]. We thus can focus our attention to selected aspects which are of importance to the topic of this review.

In most cases, cyclopropene and its derivatives are easily prepared [15]. Especially the 3,3-disubstituted cyclopropenes can be obtained on a multigram scale in a two-step process [24–26] with overall yields as high as 80% (Eq. 1)

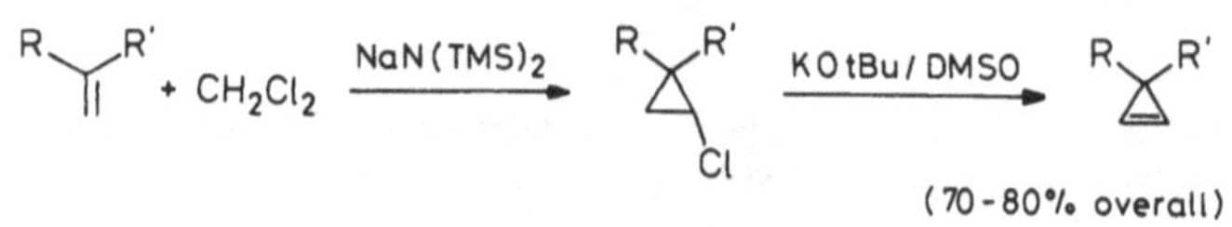

(1)

R:	CH_3,	CH_3,	C_6H_5,	$(CH_2)_4$,	CH_3, C_2H_5
R':	CH_3,	C_6H_5,	C_6H_5,		C_2H_5, C_2H_5

For the preparation of the higher boiling cyclopropenes dehydrohalogenation of the monobromocyclopropanes [27] at ca 40 °C has been found to be the most convenient method [28].

Cyclopropenes are highly strained and thus very reactive molecules. Cyclopropene itself, first reported [29] in 1922, is a potentially explosive gas (b.p. —36 °C) which tends to polymerize even below 0 °C. 1-Methylcyclopropene (b.p. 8 °C) dimerizes within minutes at room temperature via an ene reaction [30, 31] (Eq. 2), but can be stored at —78 °C for several weeks.

(2)

In contrast, 3,3-dimethylcyclopropene (b.p. 14 °C), which cannot undergo an ene reaction, can be heated in a sealed tube for several days at 100 °C without significant decomposition [32].

Introduction of an internal double bond into the cyclopropane ring increases the strain of the σ-frame work. The calculated strain energy of cyclopropene is 228 kJ/mol [33]. A main factor is increased angular strain, as can be seen by looking at the bond angles: C(1)–C(3)–C(2) 50.4°, C(3)–C(1)–C(2) and C(3)–C(2)–C(1) 64.8° (cf. Fig. 2).

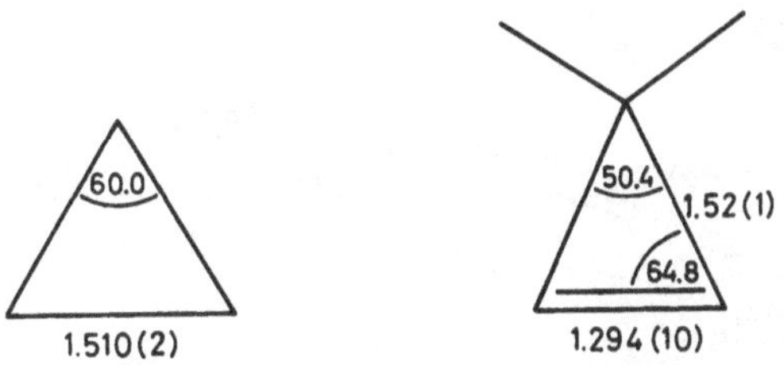

Fig. 2. Structural data of cyclopropane [34] and 3,3-dimethylcyclopropene [35] as established by electron diffraction and microwave spectroscopy, respectively

Thus the molecular σ-frame work is one source of destabilization and release of strain should be a driving force in the reactions of cyclopropenes.

Cyclopropenes are known to be very reactive dienophiles [36]. Cycloaddition across the double bond reduces the ring strain by 109 kJ/mol [37]. In the gas phase (350 °C) cyclopropene rearranges to propyne. Calculations [38] dealing with this thermal ring-opening reaction suggest that the activation energy lies between 159 and 176 kJ/mol. The lowest-energy pathway involves simultaneous ring-opening and methylene rotation to give a planar 1,3-diradical state which then can decay to the singlet and triplet carbene states with the latter being more stable by 50 kJ/mol (Eq. 3).

(3)

This ring cleavage is, in fact, reversible [39], and many cyclopropenes have been synthesized via vinylcarbenes generated from various precursors.

In order to make use of the (potential) multifunctionality of cyclopropenes (cf. Fig. 1) in synthesis, their reactivities have to be tuned. All efforts to achieve this goal finally focussed on the use of transition metal catalysts. Before we discuss the various transition metal catalyzed reactions of cyclopropenes in detail, we will briefly summarize their stoichiometric reactions with transition metal complexes (Scheme 1). These may be regarded as equivalents of important elementary steps in a catalytic cycle. Four different types of reactions have been observed. Reactions at the double bond involve

(1) π-complexation and
(2) oxidative addition. Ring-opening can occur between C^1 and C^3 as well as between C^1 and C^2. π-Complexation of a cyclopropene double bond is accompanied by release of ring strain, as can be seen by looking at the bond angles in $(PPh_3)_2Pt(\eta^2$-1,2-dimethyl-cyclopropene) (*1*) (cf. Fig. 3).

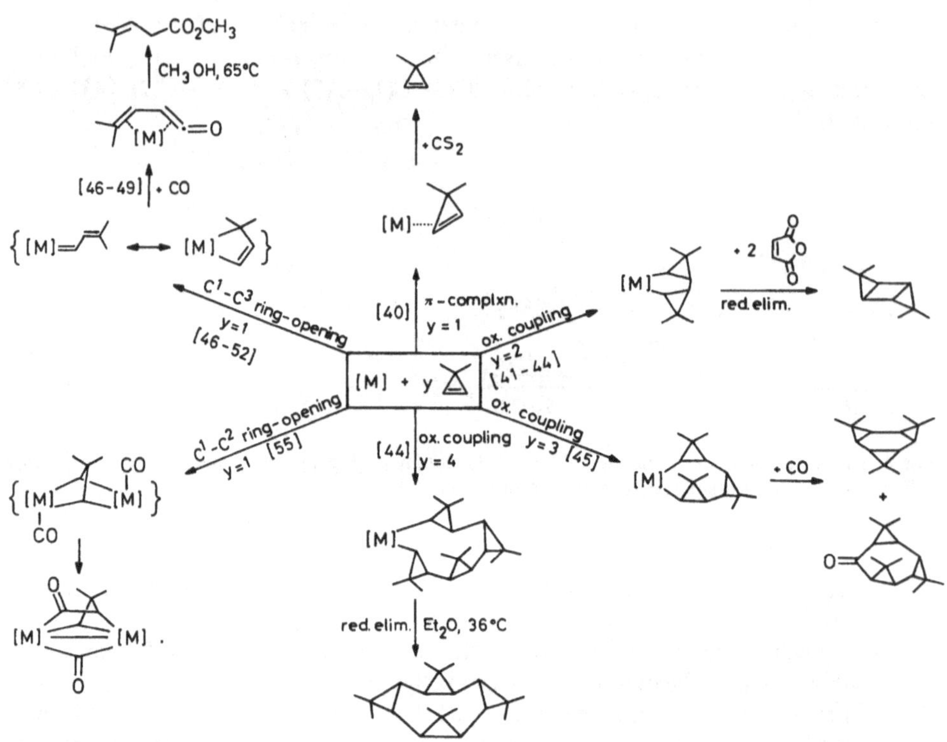

Scheme 1. Selected stoichiometric reactions of 3,3-dimethylcyclopropene with transition metal complexes

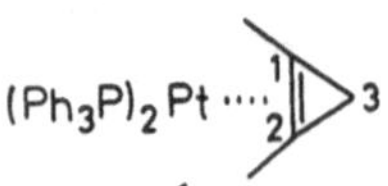

1

Fig. 3. Selected structural data of compound *1* as established by a three-dimensional X-ray analysis [40]: C(1)–C(2) 1.50(1), C(1)–C(3) 1.55(2), C(2)—C(3) 1.54(1) Å, C(2)—C(1)–C(3) 60.9(8)°, C(1)–C(3)–C(2) 58.1(7)°

These values are relatively close to the ideal value of 60°. The C=C double bond is lengthened to 1.50(1) Å. The complexation is reversible and the cyclopropene can be recovered. Therefore, such complexes have been propagated as cyclopropene storages [40]. The second important type of reaction is that of oxidative coupling. Two, three or four cyclopropene units couple with the metal to give metallacycloalkanes of different ring size. An interesting structural feature of these metallacycles are alternating C—C bond lengths within the metallacyclic ring. As an example, selected structural data of Bis(dimethylphenylphosphane)-3,3,6,6,9,9,13,13-octamethyl-11-pallada-*anti,-syn,anti,syn*-pentacyclo-[10.1.0.0.^{2,4}0.^{5,7}0^{8,10}]tridecane (*2*) are given in Fig. 4. The C—C distances within the cyclopropyl rings are 0.3 Å longer than the remaing C—C bonds within the metallcyclic ring. This is probably due to conjugative effect induced

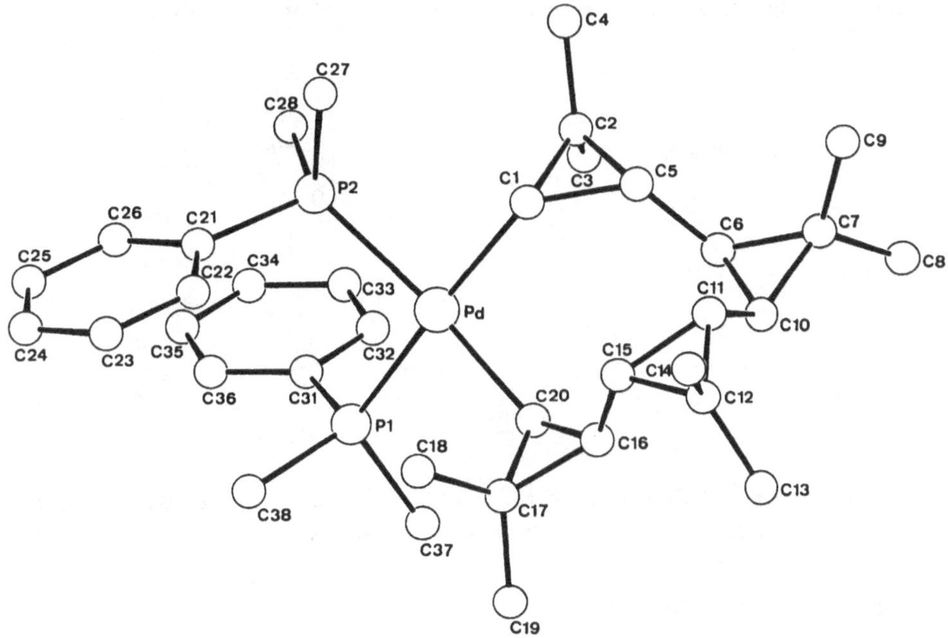

Fig. 4. Selected structural data [53] of compound *2*: Bond lengths Å: Pd—C(1) 2.085(3), Pd—C(20) 2.083(3), C(1)–C(5) 1.527(4), C(5)–C(6) 1.494(4), C(6)–C(10) 1.522(4), C(10)–C(11) 1.490(4), C(11)–C(15) 1.524(4), C(15)–C(16) 1.491(4), C(16)–C(20) 1.520(4). Bond angles [°]: C(2)–C(1)–C(5) 59.1(2), C(1)–C(2)–C(5) 60.1(2), C(1)–C(5)–C(2) 60.8(2)

by the unsaturated character of the cyclopropyl groups. The bond angles within the three-membered rings are very close to the ideal 60° values.

All these metallacycles undergo reductive eliminations under different conditions to give the respective cyclic hydrocarbons. The role of metallacycloalkanes as catalytic intermediates has recently been discussed elsewhere [54].

Ring-opening between C^1 and C^3 has been observed in reactions of cyclopropenes with iron [46–49], manganese [46, 50], and molybdenum [51, 52]. Vinylcarbene complexes or metallacyclobutenes as the primary reaction products have not yet been isolated but could be "trapped" e.g. with CO to give vinylketene complexes [46–49] or by formation of an allyl-bridged dimolybdenum complex [51]. However, with 1,2-diphenylcyclopropenone a platinacyclobutenone *3* as a direct ring-opening product could be isolated [55] (Eq. 4).

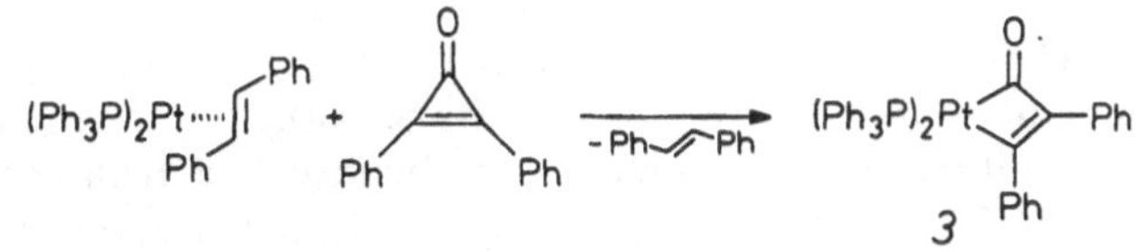

(4)

Only one example of ring-opening between C^1 and C^2 has been published so far [56] (Eq. 5).

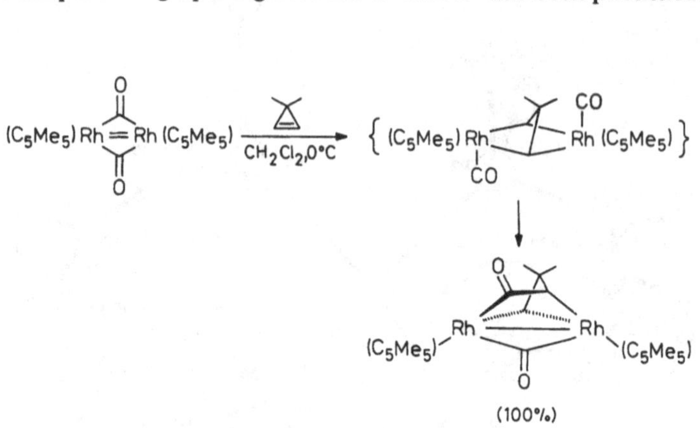

(5)

It should be noted that most of the reactions of cyclopropenes with transition metal complexes proceed at room temperature or below.

2.2 Synthesis of Cyclic Hydrocarbons

2.2.1 Three-Membered Rings

Cyclopropenes may serve as starting materials for the preparation of three-membered carbocycles in two different ways, as indicated by the arrows below:

(5a)

The first possiblity involves a stereoselective 1,2-addition to a cyclopropene double bond. The second consists of an "in-situ" generation of vinylcarbenes followed by a [2+1]-cycloaddition reaction.

1,2-Additions to the double bond of cyclopropenes have recently found considerable attention from both a synthetic and a mechanistic point of view. For example, Lehmkuhl et al. [57] have developed a stereoselective synthesis of chrysanthemic acid via a 1,2-Grignard addition to 3,3-dimethylcyclopropene. Nesmeyanova et al. [58] have found that the bromination of 3,3-dimethylcyclopropene possibly proceeds via a concerted, electrocyclic trans-addition.

Transition metal catalysts allow stereoeselective *cis*-1,2-additions without having to deal with undesired byproducts such as magnesium salts. As an example, stereoselective functionalization of 3,3-dimethylcyclopropene can be achieved via cotrimerization with methyl acrylate in the presence of ligand-modified nickel(0) catalysts, to give the cotrimers *4*, *5* and *6* in high yields [59] (Eq. 6).

Yields and product distribution strongly depend on the nature of the ligand L where L is a phosphane or phosphite (cf. Table 1). With the bulky phosphane $P(iPr)_2(tBu)$ only 6% of the 1,2-addition products *4* and *5* are obtained, the main products being 2:1 cyclocotrimers *6* of 3,3-dimethylcyclopropene and methyl acrylate (see p. 85).

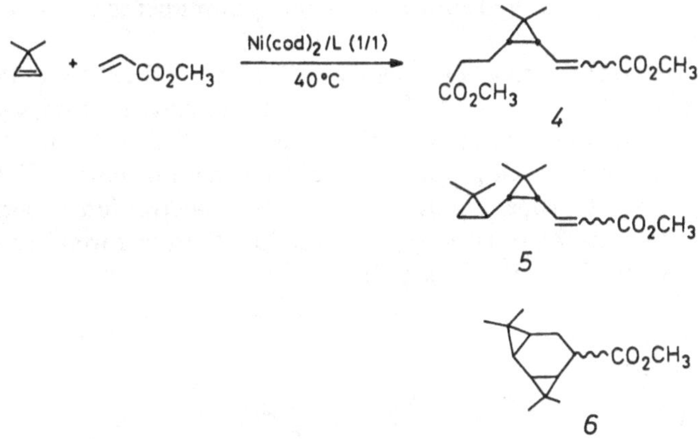

(6)

This result may reflect a hampered β—H elimination in the organonickel intermediates. With other phosphanes or phosphites the yields of *4* and *5* range from 56 to 86%. The data collected in Table 1 reveal that there is no simple correlation between product distribution and basicity or cone angle [60] of L.

Table 1. L—Ni(0) catalyzed cotrimerization of 3,3-dimethylcyclopropene with methyl acrylate: dependence of the product distribution on L

L	χ_{Tol} [cm^{-4}]	θ_{Tol} [°]	yields (%)			
			4	*5*	*6*	Σ
$P(C_2H_5)_3$	5.6	132	76	10	6	92
$P(iPr)_3$	3.1	160	12	50	17	79
$P(c\text{-}C_6H_{11})_3$	0.3	172	13	66	6	85
$P(iPr)_2(tBu)$	2.0	167	3	3	67	73
$P(CH_2-C_6H_5)_3$	10.3	160	48	13	32	93
$P(C_6H_5)_2CH_3$	11.2	136	49	12	19	80
$P(C_6H_6)_3$	12.9	145	37	26	17	80
$P(O-C_2H_5)_3$	20.4	109	52	17	31	100
$P(O-iPr)_3$	19.8	130	51	5	32	88
$P(O-C_6H_5)_3$	29.2	128	78	3	10	91
$P(O\text{-}o\text{-}C_6H_4-C_6H_5)_3$	28.9	152	48	9	22	79

Franck-Neumann et al. [61] have succeeded in preparing *cis*-chrysanthemic acid methyl ester (*8*) by selective catalytic hydrogenation of the corresponding cyclopropene derivative *7* (Eq. 7).

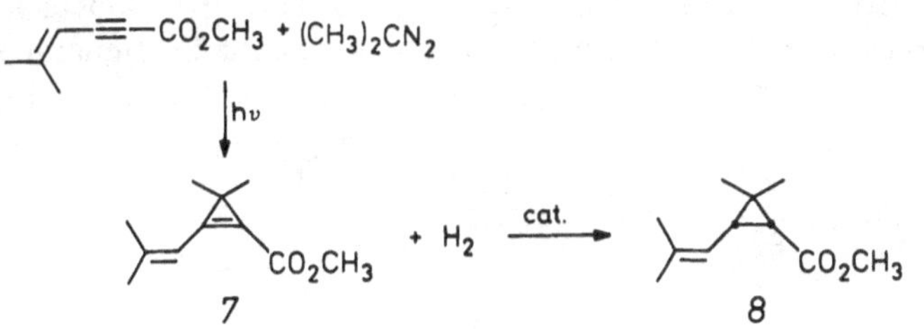

(7)

The rhodium(I) catalyzed hydroformylation of cyclopropenes to give *cis*-aldehydes has been mentioned [62].

Among the methods at hand to synthesize cyclopropane derivatives, carbene addition to alkenes plays a prominent role [63]. As a source of vinylcarbenes, cyclopropenes might be useful in this kind of approach. In 1963, Stechl was the first to observe a transition metal catalyzed cyclopropene-vinylcarbene rearrangement [64]. When treating 1,3,3-trimethylcyclopropene with copper salts, dimerization occurred to give 2,3,6,7-tetramethyl-octa-2,4,6-triene (*9*), the product from a formal recombination of the corresponding vinylcarbene (Eq. 8).

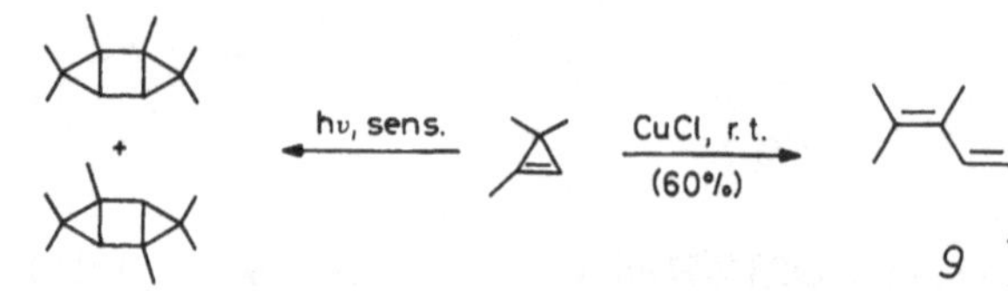

9

(8)

In the meantime thermal [65] and metal catalyzed [66,67] rearrangements of cyclopropenes have been detected as convenient methods for the preparation of vinylcyclopropanes via formal [2+1]-cycloadditions of vinylcarbenes to alkenes (Eq. 9) (for an alternative entrance starting from allylidene dichloride or 1,3-dichloropropene, see Ref. [68]).

(9)

Some thermally unstable cyclopropenes are known, which react with olefines at or below ambient temperature by ring-opening to give vinylcyclopropanes (Eq. 10) [65]. Cyclopropenone ketals react with electron deficient olefines in the same manner, but here an ionic mechanism is more likely [65e].

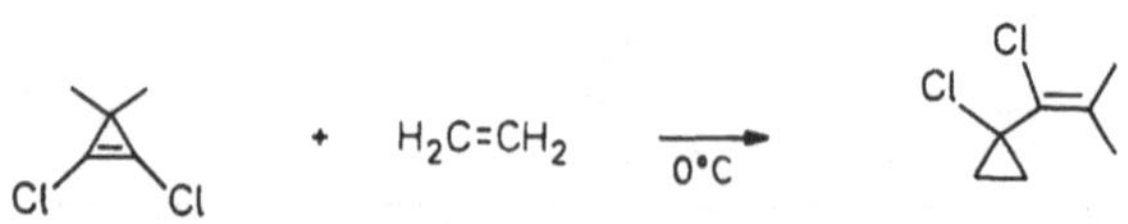

(10)

The thermally more stable alkyl- or arylsubstituted cyclopropenes can undergo this reaction type with the aid of transition metal catalysts under mild conditions. The choice of a suitable catalyst strongly depends on the nature of the olefinic cosubstrate. For electron-deficient alkenes, $Ni(cod)_2$ (where cod = *cis* cycloocta-1,5-diene) has been found to be the best catalyst [66]. Dialkyl maleates, dialkyl fumarates and methyl

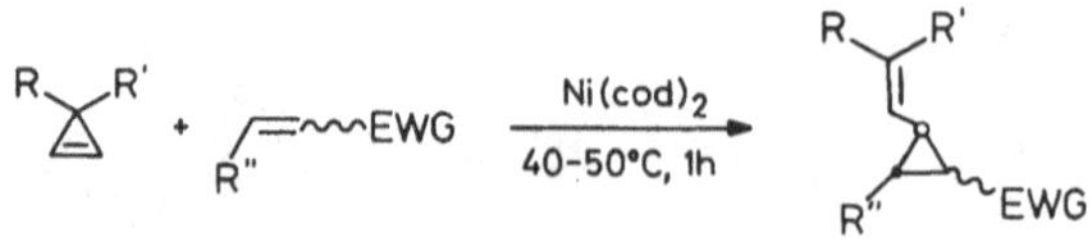

(40–73%)

(11)

acrylate react smoothly with 3,3-diorganylcyclopropenes to give the corresponding vinylcyclopropane derivatives (Eq. 11).

The isolated yields range from 40 to 73%, the byproducts being isomers and cyclodimers of the corresponding cyclopropenes as well as cotrimers. These byproducts are easily separated by fractional distillation. The stereochemical course of the codimerization of 3,3-dimethylcyclopropene with dimethyl maleate (Eq. 12) provides some useful information on the nature of the reaction intermediates.

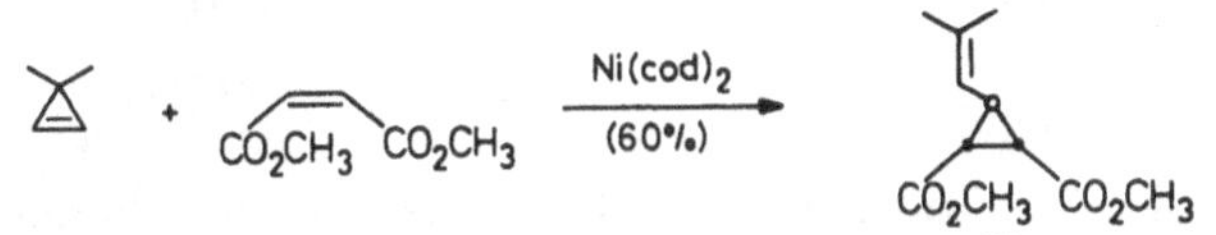

10 (12)

In a large number of carbene and carbenoid addition reactions to alkenes the thermodynamically less favored *syn*-isomers are formed [63]. The finding that in the above cyclopropanation reaction the *anti*-isomer is the only product strongly indicates that the intermediates are organonickel species rather than carbenes or carbenoids. Introduction of alkyl groups in the 3-position of the electron-deficient alkene hampers the codimerization and favors isomerization and/or cyclodimerization of the cyclopropenes. Thus, with methyl crotylate and 3,3-diphenylcyclopropene only 16% of the corresponding vinylcyclopropane derivative has been obtained. 2,2-Dimethyl acrylate does not react at all with 3,3-dimethylcyclopropene to afford *trans*-chrysanthemic acid methyl ester. This is in accordance with chemical expectations [69] since in most cases the tendency of alkenes to coordinate to Ni(0) decreases in the order un-, mono- $<$ di- $\ll$ tri- $<$ tetrasubstituted olefines.

For the reaction of olefinic cosubstrates without activating groups copper(I) catalysts have to be used (Eq. 13).

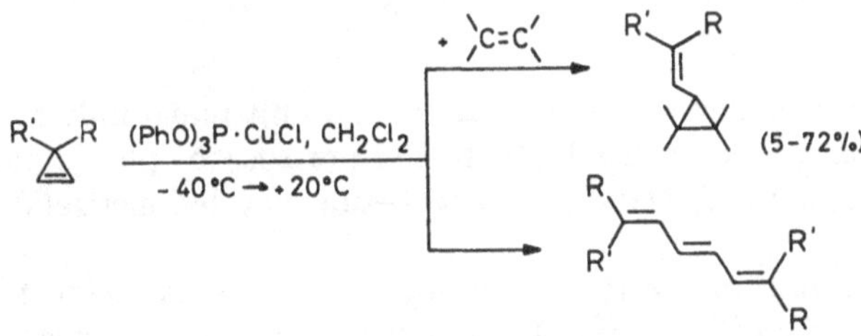

(13)

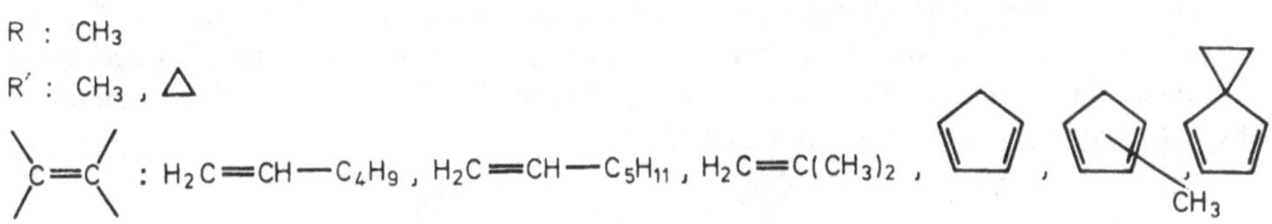

With hexene-1 and heptene-1, the yields of vinylcyclopropane derivatives are rather low (5–10%), the main products being the corresponding hexatriene derivatives. With the other alkenes listed above, the yields of cyclopropanation products range from 55 to 77% [67].

An interesting finding is the predominant formation of *syn*-isomers by monocyclopropanation of cyclopentadienes [67b)]. This is what one would expect when carbenes or carbenoids are involved [63)]. Also one isomer is found in the codimerization of 1-propyl-cyclopropene-3-carboxylate (*11*) with norbornadiene (Eq. 14). .

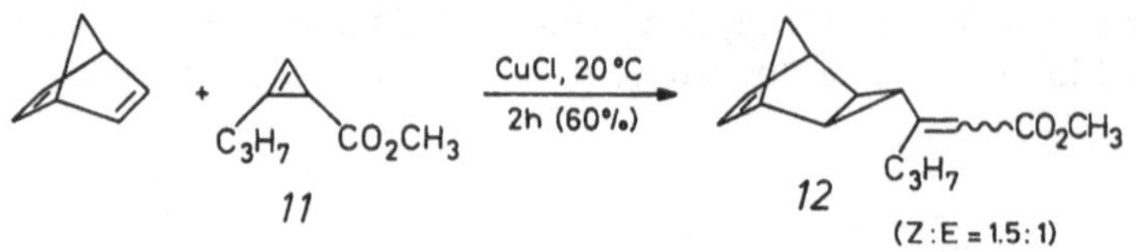

(14)

In the presence of CuCl, exo-3-tricyclo[3.2.1.0^{2,4}]octene-6-yl-3-hexene-2-oic acid methyl ester (*12*) is isolated after TLC in 60% yield ($Z:E = 1.5:1$) [67c)]. This is comparable to the course of the cyclopropanation of norbornadiene with both the Simmons-Smith reagent and diazomethane-CuCl. In both cases the *exo-anti* route is favored over the *exo-syn* [63)]. It is important to notice, that in the above mentioned nickel(0) catalyzed reactions (Eq. 11) hexatriene derivatives have never been observed.

2.2.2 Four-Membered Rings

It has already been mentioned that cycloaddition across the cyclopropene double bond reduces the ring strain by about 109 kJ/mol. [2+2] cyclo- and -codimerizations of cyclopropenes should therefore be thermodynamically favored processes (see also Sect. 3.2). Indeed, when catalyzed by Zeolites [70)], Lewis acids [71)] and transition metal complexes [30, 72–74)] *anti*-tricyclo[3.1.0.0^{2,4}]hexane derivatives can be synthesized efficiently (Eq. 15).

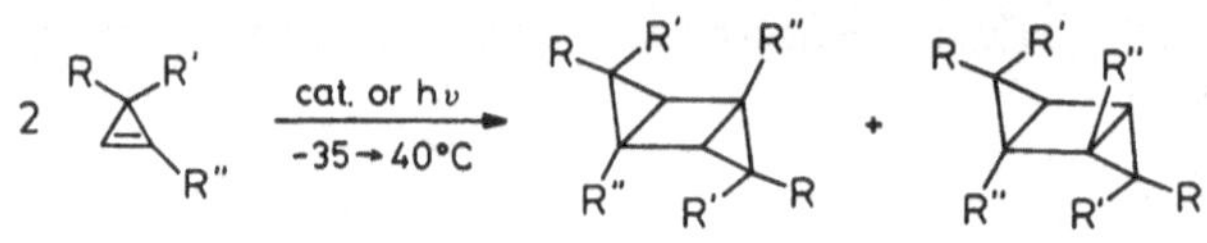

(15)

In most cases, the cyclodimerizations proceed smoothly under mild conditions and in good to excellent yields (cf. Table 2). In the presence of phosphane-free Pd(0) catalysts some cyclotetramerization occurs besides cyclodimerization [73, 74)] (see p. 97).

The *anti*-configuration of the tricyclohexane derivatives has been deduced from NMR-spectroscopic data [71)] and may be compared with the molecular geometry of a photodimer of methyl 1,2-diphenylcyclopropene-3-carboxylate that has been determined by X-ray analysis [75)]. A (bridged) *syn*-tricyclohexane derivative *14* has been obtained via intramolecular [2+2]-cycloaddition [76)] of the destabilized diene *13* (Eq. 16), catalyzed by a metathesis catalyst.

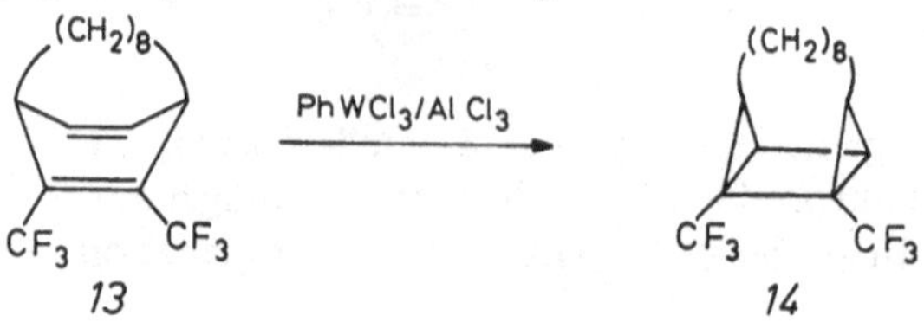

(16)

Table 2. Summary of catalytic [2+2]-cyclodimerization reactions of cyclopropenes (cf. Eq. 15)

entry	catalyst	R	R′	R″	yield (%)	Ref.
1	zeolite	H	H	H	97	[70]
2	zeolite	CH_3	H	H	96	[70]
3	zeolite	H	H	CH_3	n.d.	[70]
4	BF_3OEt_3	CH_3	CH_3	H	83	[71]
5	BF_3OEt_2	CH_3	C_6H_5	H	31	[72]
6	BF_3OEt_2	C_2H_5	C_2H_5	H	70	[72]
7	BEt_3	CH_3	CH_3	H	58	[71]
8	$AlCl_3OEt_2$	CH_3	CH_3	H	91	[71]
9	$AlEt_3OEt_2$	CH_3	CH_3	H	81	[71]
10	$(\eta^5\text{-}C_5H_5)Co(cod)$	CH_3	CH_3	H	40	[72]
11	"$Pd(dba)_2$"	CH_3	CH_3	H	80	[73]
12	"$Pd(dba)_2$"	CH_3	C_2H_5	H	82	[74]
13	"$Pd(dba)_2$"	C_2H_5	C_2H_5	H	71	[74]
14	"$Pd(dba)_2$"	$-(CH_2)_4^-$		H	53	[74]
15	$PdCl_2$	H	H	CH_3	ca. 23	[30]
16	$[(\eta^3\text{-}C_3H_5)PdCl]_2$	CH_3	CH_3	CH_3	10	[30]

cod = *cis*-cycloocta-1,5-diene
dba = dibenzylideneacetone

1,2-Disubstituted cyclopropenes do not react in the above sence. 1,2-dimethylcyclopropene polymerizes at 0 °C in the presence of $[(\eta^3\text{-}C_3H_5)PdCl]_2$ [30]. 1,2-Diphenylcyclopropene cyclodimerizes nearly quantitatively in the presence of "$Pd(dba)_2$" or $Pd(\eta^5\text{-}C_5H_5)(\eta^3\text{-}C_3H_5)$ to yield 1,2,4,5-tetraphenyl-cyclohexa-1,4-diene [72] (see p. 96).

Catalytic codimerizations between cyclopropenes and alkenes (or alkynes) to give four-membered rings have only been achieved with 3,3-dimethyl- and 3-cyclopropyl-3-methylcyclopropene on the one hand and norbornene or norbornadiene on the other (stereochemistry not determined) [77]:

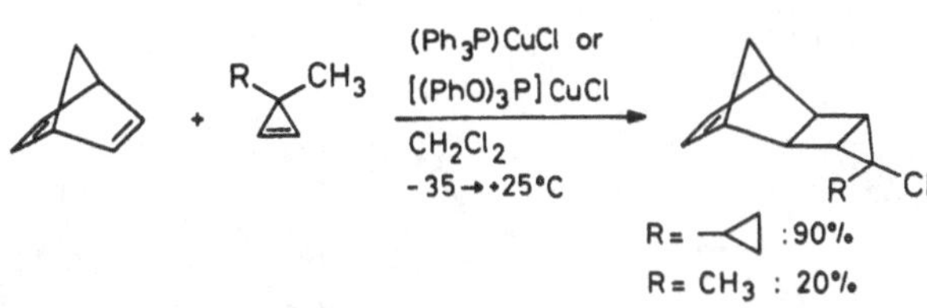

(17)

1-Methylcyclopropene is reported to react with activated alkynes at —30 °C via an ene reaction to give vinylcyclopropene derivatives [30]. In the presence of silver perchlorate, bicyclopropenyl derivatives such as *15* isomerize to bicyclo[2.2.0]hexa-2,5-diene derivatives (e.g. *16*), which can be isolated [78,79].

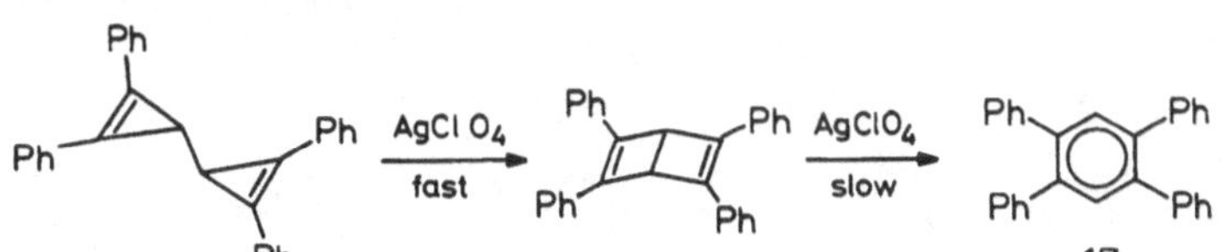

15 *16* *17* (18)

These rearrangements involve cationic argentovinylcarbene intermediates whereas in the case of the Pd-catalyzed cyclodimerization reactions the intermediacy of palladacyclopentanes has been proven (Scheme 1).

The *trans*-tricyclo[3.1.0.0^{2,4}]hexane derivatives, prepared according Eq. 15, are easily isomerized by some Rh(I) catalysts to give *trans*-1,2- and 1,3-divinylcyclobutane derivatives [72] (Eq. 19).

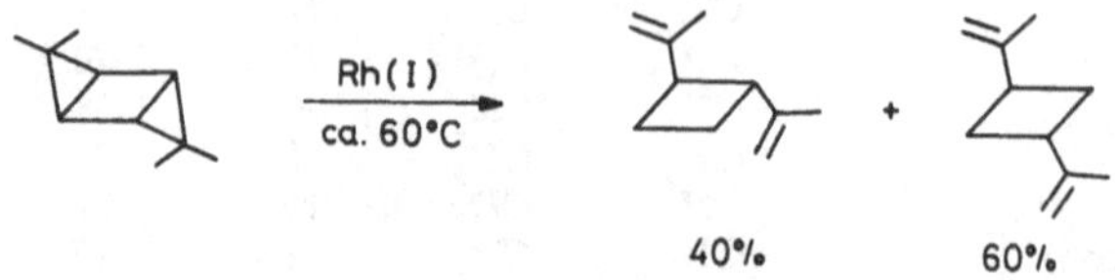

(19)

Rh(I) = $CpRh(CH_2{=}CH_2)_2$; $[(CO)_2RhCl]_2$

In contrast to these results the catalytic hydrogenation (PtO_2/glacial acid) of these cyclodimers occurs by cleavage of the bridging σ-bonds of the three-membered ring leading to cyclohexane derivatives in high yield (Eq. 20). These results are consistent with the rules of Musso [80a)], but unexpectedly 3,3-dimethylbicyclo[2.1.0]pentane is hydrogenated by cleavage of the cyclopropane ring in the neighborhood of the gem. dimethyl group [80b)].

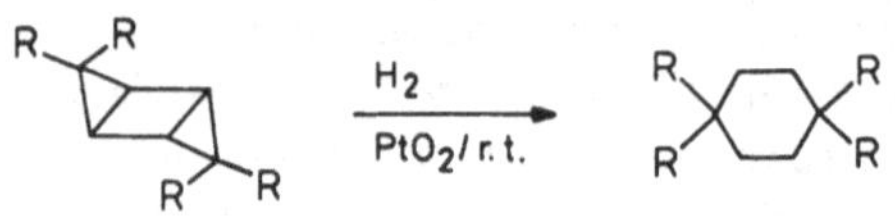

(20)

2.2.3 Five-Membered Rings

In recent years, cyclopropenes have been used successfully as starting materials for the preparation of five-membered carbocycles. Three different approaches may be envisaged:

1) isomerization of 3-substituted cyclopropenes (Eq. 21)
2) [2+2+1]-cycloadditions (Eq. 22)
3) [3+2]-cycloadditions (Eq. 23).

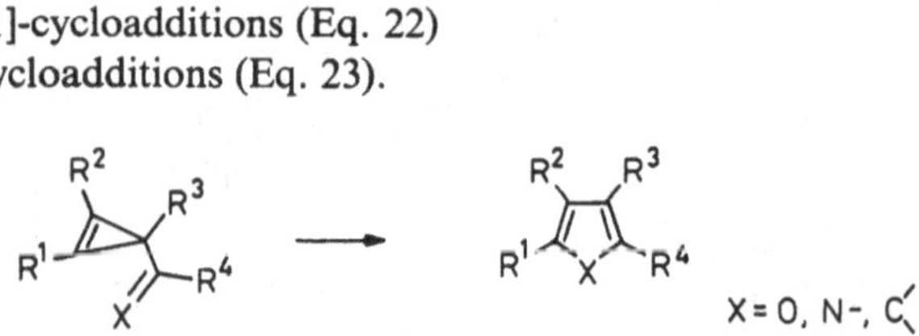

(21)

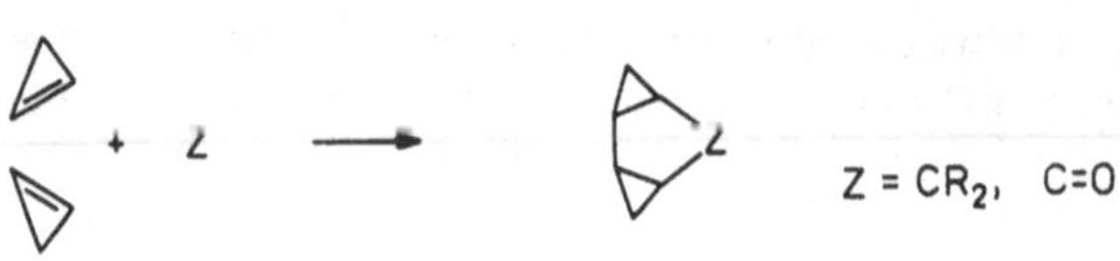

(22)

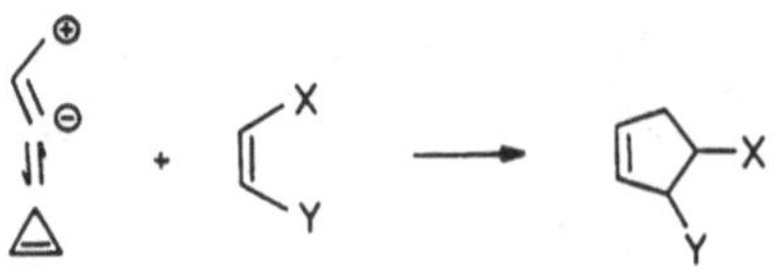

(23)

The first approach has been realized in various ways. The second one only works with the aid of transition metal catalysts, whereas the third one has been realized so far only in the thermal reaction of cyclopropenone acetals with electron-deficient alkenes [81].

In 1966, Komendantov et al. [82] were the first to observe a copper(I) catalyzed quantitative isomerization of ethyl-1,2-dipropylcyclopropene-3-carboxylate to 2,3-dipropyl-5-ethoxyfurane. This reaction principle has subsequently been extended to a number of substituted cyclopropenes [83–85] (Eq. 24).

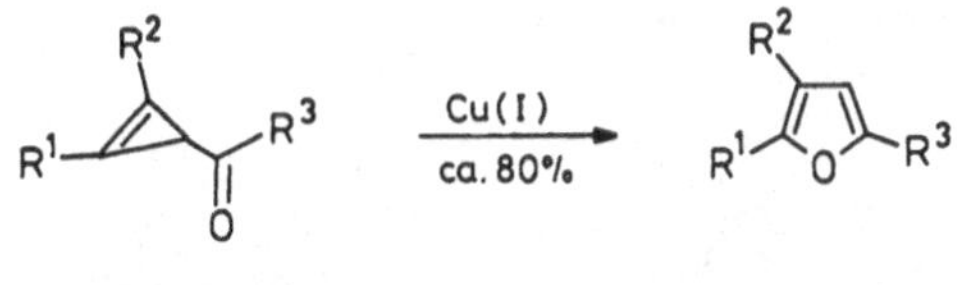

(24)

R¹: n-C_3H_7, C_6H_5, n-C_4H_9, t-C_4H_9
R²: CH_3, n-C_3H_7, C_6H_5
R³: H, CH_3, $CH_2C_6H_5$, OC_2H_5
Cu(I): copper stearate

A bipyrrole *19* has also been obtained by treating the corresponding aldazine *18* with catalytic amounts of copper stearate [83] (Eq. 25).

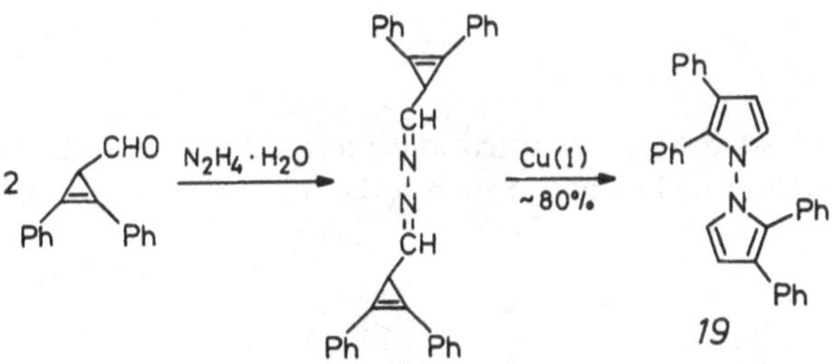

18 *19*

(25)

With 2,3-diphenylcycloprop-2-ene-1-carboxylic acid (*20*) a 90% yield of 4,5-diphenyl-2(3 H)-furanone (*21*) is obtained [86] (Eq. 26).

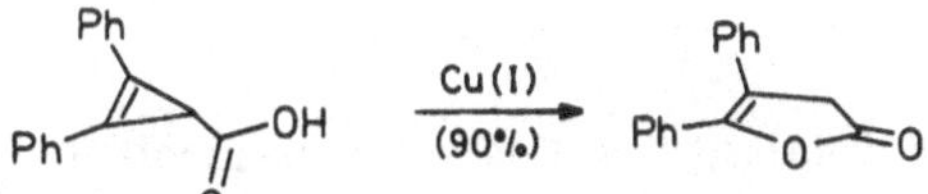

20 *21*

(26)

The mechanism of these copper catalyzed isomerizations is not fully understood. A concerted sigmatropic 1,3-shift as well as a biradical process have been proposed but a decision between these two possibilities could not be made [85]. Kinetic measurements at least have shown that the reactions are not monomolecular and that the reaction rates depend on the catalyst concentration.

Cyclopropenes bearing a vinyl or an aryl group in the 3-position easily undergo cyclopropene — cyclopentadiene (Eq. 28) and cyclopropene — indene rearrangements (Eq. 29), respectively. These reactions are formally analogeous to the vinylcyclopropane — cyclopentene rearrangement (Eq. 27), but the intermediancy of a vinylcarbene has been proven in one case [65d].

(27)

(28)

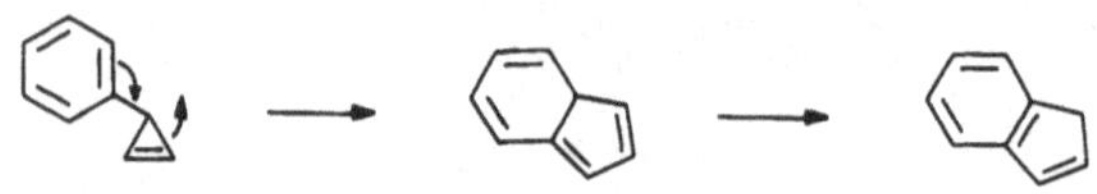

(29)

The cyclopropene — indene rearrangement is a well known reaction which can be affected by acid [87], heat [88, 65d] or irradiation [20]. The first transition metal catalyzed isomerization [89] has been published in 1968 (Eq. 30).

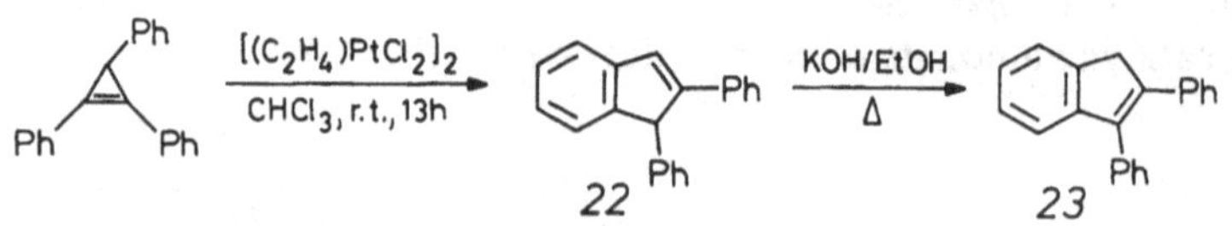

(30)

A number of catalysts have been tested and literatur survey up to 1980 has been published. Silver perchlorate has been found to be most effective [90] (Eq. 31).

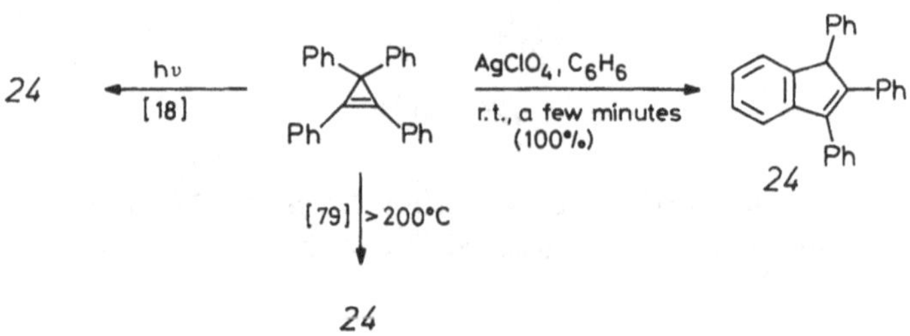

(31)

Another interesting class of catalysts for this rearrangement are palladium(0) complexes, e.g. "$Pd(dba)_2$". These complexes also catalyze the isomerization of vinylcyclopropenes to cyclopentadienes [72] (Eq. 33).

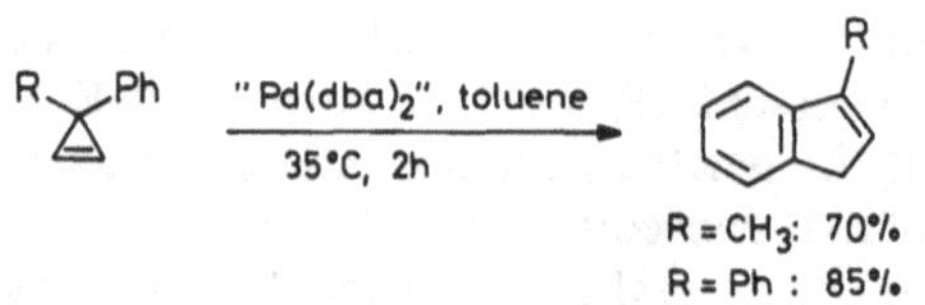

(32)

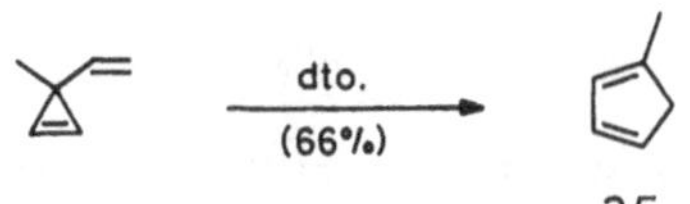

25

(33)

The [2+2+1]-cycloaddition approach to five-membered carbocycles has been realized in the case of the nickel(0) catalyzed co-trimerization of 3,3-dimethylcyclopropene with CO [91] (Eq. 34).

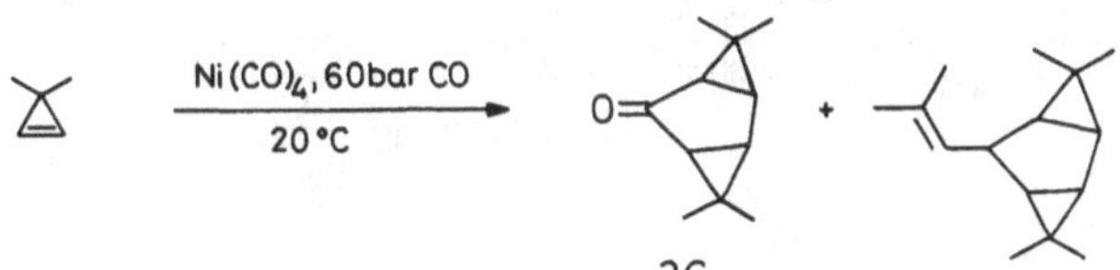

26 (54%) 27 (24%)

(34)

In the presence of catalytic amounts of $Ni(CO)_4$ the tricyclic derivatives *26* and *27* are obtained in 54% and 24% yield, respectively, when 3,3-dimethylcyclopropene is treated with excess CO at room temperature. The *anti*-arrangement of the three-membered rings in *26* and *27* has been established by ^{1}H-NMR spectroscopy [91]. Equation 34 represents the first example of a catalytic [2+2+1]-cycloaddition reaction of two alkenes with carbon monoxide or a carbene unit, respectively [92]. Other cosubstrates than CO have not yet been examined.

2.2.4 Six-Membered Rings

Two strategies are conceivable to transform cyclopropenes into six-membered carbocycles:

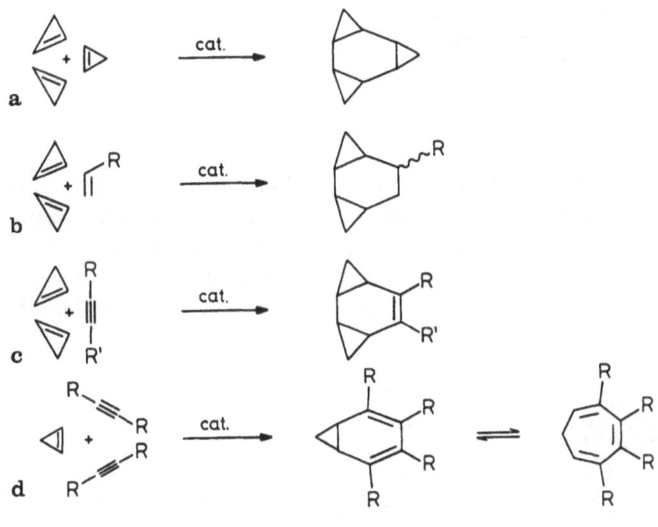

(35)

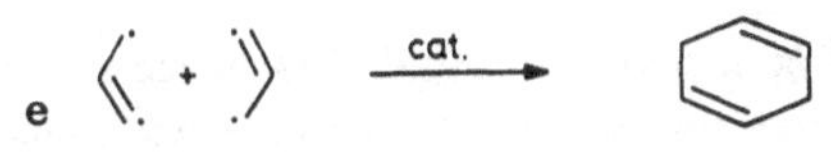

(36)

1) [2+2+2]-cycloadditions (Eq. 35)
2) [3+3]-cycloaddition (Eq. 36)

In theory, electrocyclic [2+2+2]-cycloadditions should be thermally allowed [93] and they are in most cases exothermic reactions [94]. Nevertheless, only a few examples of thermally induced [2+2+2]-cycloadditions are known [94], the most prominent example probably being the codimerization of norbornadiene with tetracyanoethylene [95a] (Eq. 37).

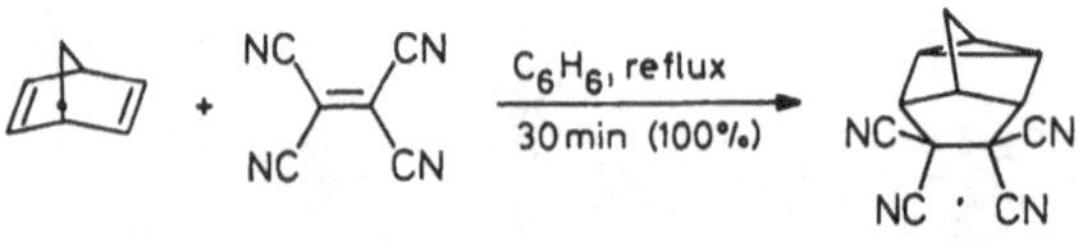

28 (37)

Compared with the Diels-Alder reaction, the [2+2+2]-cycloaddition is potentially more powerful since the number of new bonds as well as chirality centers that are formed is higher. Unfortunately, the reaction seems to be entropically or kinetically unfavorable. This disadvantage can, however, be overcome by the use of transition metal catalysts (templates). Among the most successful examples of this reaction type, the nickel(II) catalyzed Reppe reactions [96], the cobalt(I) catalyzed cocyclizations of α,ω-diynes with alkynes [97], the cobalt(I) catalyzed pyridine synthesis [98] and last but not least the palladium(0) catalyzed cyclotrimerizations of 3,3-dialkylcyclopropenes to *trans*-σ-tris-homobenzenes must be mentioned. The latter has been known for ten years [99].

In the presence of trialkylphosphane-modified Pd(0), 3,3-dimethylcyclopropene is cyclotrimerized quantitatively and stereoselectively to give 3,3,6,6,9,9-hexamethyl-*cis,cis,trans*-tetracyclo[6.1.0.$0^{2,4}0^{5,7}$]nonane (*29*) [73, 99] (Eq. 38).

29 (38)

This method has been extended to other 3,3-disubstituted cyclopropenes [74]. The structure of *29* which has primarily been derived from spectroscopic data [99], has been confirmed by a three-dimensional X-ray analysis [100]. The two *syn*-oriented dimethylcyclopropyl rings impose steric strain on the molecule which faciliates the thermal rearrangement to hexamethyl-trans-tricyclo[4.3.0.$0^{7,9}$]non-3-ene (*30*) [101] (Eq. 39)..

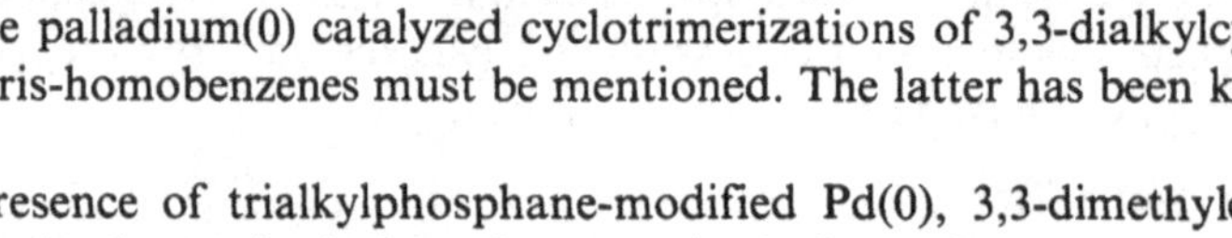

29 30 31 (39)

It has been shown by ^{12}C labelling experiments that this process proceeds via a $[\sigma_s2 + \sigma_s2 + \sigma_s2]$-cycloreversion to give the corresponding *cis,trans,trans*-cyclonona-

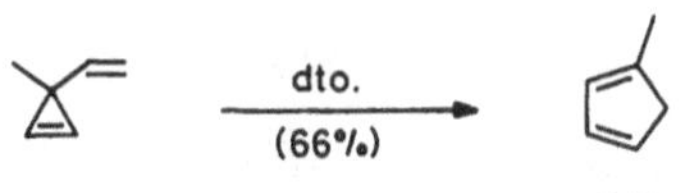

(33)

25

The [2+2+1]-cycloaddition approach to five-membered carbocycles has been realized in the case of the nickel(0) catalyzed co-trimerization of 3,3-dimethylcyclopropene with CO [91] (Eq. 34).

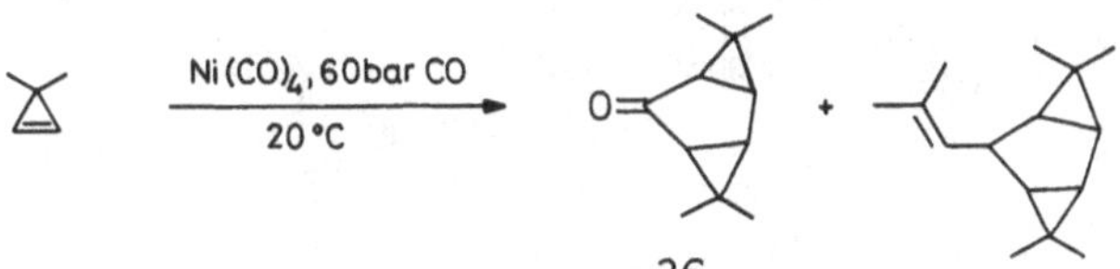

(34)

26 (54%) 27 (24%)

In the presence of catalytic amounts of $Ni(CO)_4$ the tricyclic derivatives *26* and *27* are obtained in 54% and 24% yield, respectively, when 3,3-dimethylcyclopropene is treated with excess CO at room temperature. The *anti*-arrangement of the three-membered rings in *26* and *27* has been established by ^{1}H-NMR spectroscopy [91]. Equation 34 represents the first example of a catalytic [2+2+1]-cycloaddition reaction of two alkenes with carbon monoxide or a carbene unit, respectively [92]. Other cosubstrates than CO have not yet been examined.

2.2.4 Six-Membered Rings

Two strategies are conceivable to transform cyclopropenes into six-membered carbocycles:

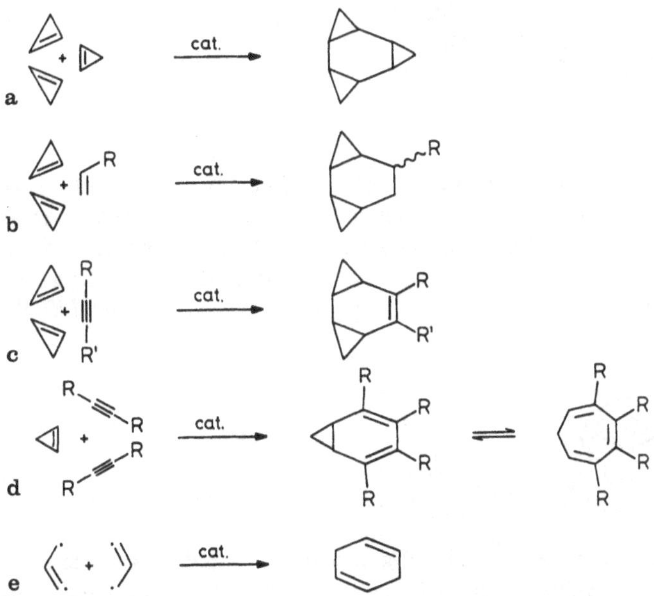

(35)

(36)

1) [2+2+2]-cycloadditions (Eq. 35)
2) [3+3]-cycloaddition (Eq. 36)

In theory, electrocyclic [2+2+2]-cycloadditions should be thermally allowed [93] and they are in most cases exothermic reactions [94]. Nevertheless, only a few examples of thermally induced [2+2+2]-cycloadditions are known [94], the most prominent example probably being the codimerization of norbornadiene with tetracyanoethylene [95a] (Eq. 37).

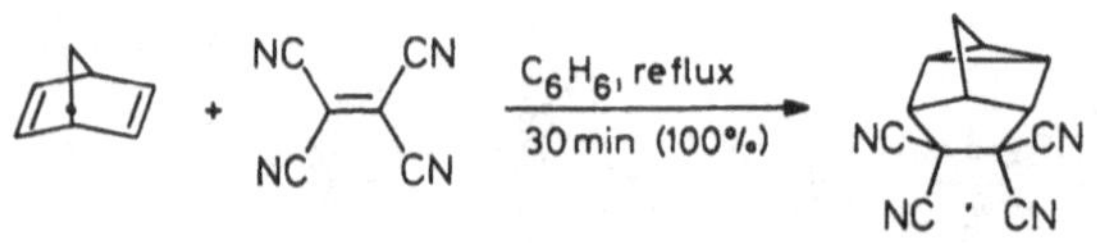

28 (37)

Compared with the Diels-Alder reaction, the [2+2+2]-cycloaddition is potentially more powerful since the number of new bonds as well as chirality centers that are formed is higher. Unfortunately, the reaction seems to be entropically or kinetically unfavorable. This disadvantage can, however, be overcome by the use of transition metal catalysts (templates). Among the most successful examples of this reaction type, the nickel(II) catalyzed Reppe reactions [96], the cobalt(I) catalyzed cocyclizations of α,ω-diynes with alkynes [97], the cobalt(I) catalyzed pyridine synthesis [98] and last but not least the palladium(0) catalyzed cyclotrimerizations of 3,3-dialkylcyclopropenes to *trans*-σ-tris-homobenzenes must be mentioned. The latter has been known for ten years [99].

In the presence of trialkylphosphane-modified Pd(0), 3,3-dimethylcyclopropene is cyclotrimerized quantitatively and stereoselectively to give 3,3,6,6,9,9-hexamethyl-*cis,cis,trans*-tetracyclo[6.1.0.0^{2,4}0^{5,7}]nonane (*29*) [73, 99] (Eq. 38).

29 (38)

This method has been extended to other 3,3-disubstituted cyclopropenes [74]. The structure of *29* which has primarily been derived from spectroscopic data [99], has been confirmed by a three-dimensional X-ray analysis [100]. The two *syn*-oriented dimethylcyclopropyl rings impose steric strain on the molecule which faciliates the thermal rearrangement to hexamethyl-trans-tricyclo[4.3.0.0^{7,9}]non-3-ene (*30*) [101] (Eq. 39)..

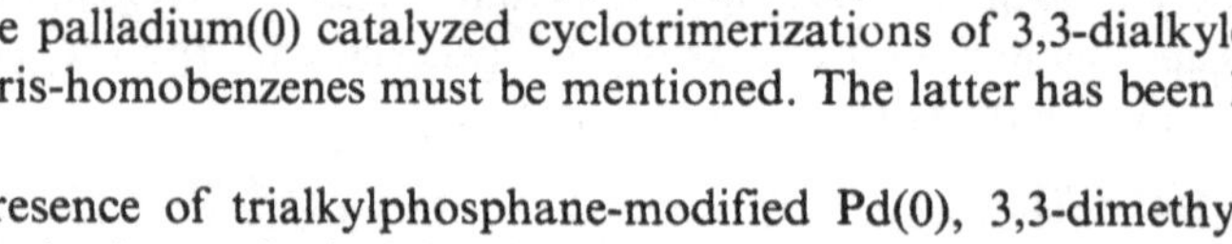

29 30 31 (39)

It has been shown by ^{12}C labelling experiments that this process proceeds via a [$\sigma_s 2 + \sigma_s 2 + \sigma_s 2$]-cycloreversion to give the corresponding *cis,trans,trans*-cyclonona-

1,4,7-triene derivative followed by a $[\pi^2s + \pi^2a]$-cycloaddition between the two strained trans double bonds [102].

In contrast to the thermal rearrangement of *29*, Rh(I) compounds catalyze both, ring-opening at the bridging σ-bond and at the other σ-bonds of the three-membered rings. Whereas a cationic Rh(I) compound catalyzes a rapid isomerization to 1,3,5-triisopropylbenzene, $CpRh(C_2H_4)_2$ induces isomerization to compounds *31a* and *31b* [72].

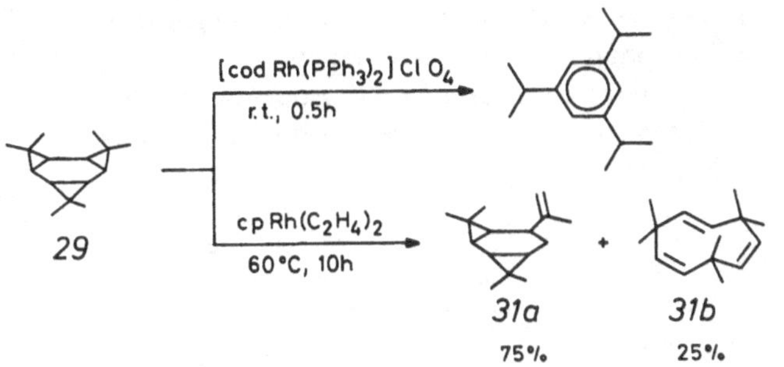

(40)

Catalytic hydrogenation of *29* on the other hand is in full agreement with the rules of Musso [80]. That means all three bridging σ-bonds of the three-membered rings are opened to yield 1,1,4,4,7,7-hexamethylcyclononane as the only product [72].

Catalytic 2:1-cotrimerizations between cyclopropenes and alkenes as well as alkynes are also known. It has already been mentioned (see Chap. 2.2.1) that a nickel(0) catalysts — modified by the bulky phosphane $P(iPr)_2(tBu)$ — favors the cocyclotrimerization of 3,3-dimethylcyclopropene and methyl acrylate [59] (Eq. 41).

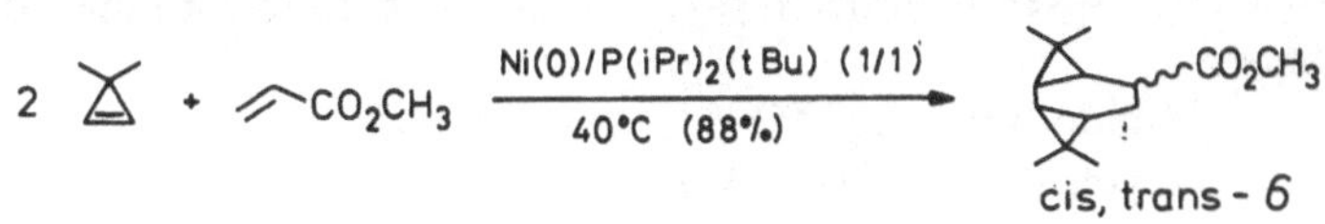

(41)

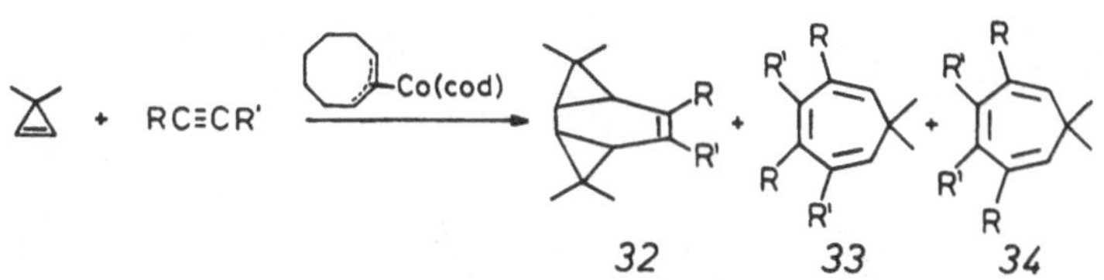

(42)

R	R'	*32*	*33*	*34*
H	C_2H_5	70%	—	—
CH_3	CH_3	84%	3%	—
H	Ph	89%	—	—
H	CO_2CH_3	67%	10%	5%
CO_2CH_3	CO_2CH_3	31%	2%	—
CH_3	$N(C_2H_5)_3$	59%	—	—

In methyl acrylate as a solvent the reaction proceeds at 40 °C to give a 88% yield of *cis-*, *trans-* *6*.

With alkynes as cosubstrates satisfactory yields of cyclotrimers are obtained in the presence of phosphane-free Co(I) catalysts, e.g: η^3-cyclooctenyl-η^4-cycloocta-diene cobalt(I). The main products are tricyclo[4.0.0^{4,6}]oct-2-ene derivatives *32*, sometimes accompanied by small amounts of the norcaradiene/cycloheptatriene derivatives *33* and *34* (Eq. 42) [27].

Since selective methods for the preparation of cycloheptatrienes are still in great demand [103] it might be worthwile to do further experiments and improve the yield of the latter compounds in the above reaction.

A formal [3+3]-cyclodimerization of cyclopropenes via ring-opening (Eq. 36) to give six-membered carbocycles has only been observed in the case of 1,2-diphenyl-cyclopropene [72] (Eq. 43).

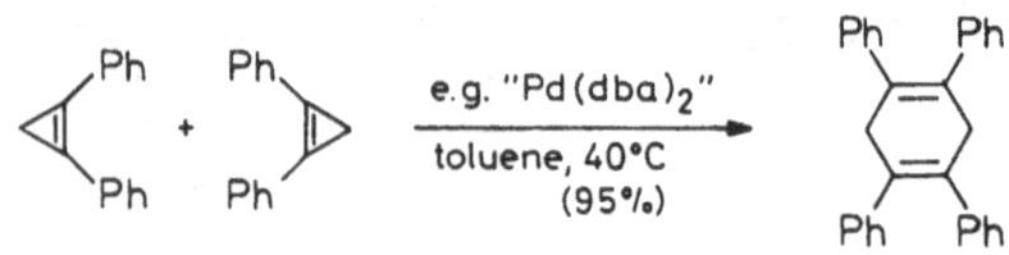

35

(43)

The reaction is catalyzed both by Ni(0) and Pd(0) catalysts, the latter giving higher yields. This transformation may occur by ring-opening as observed with some 1,2-disubstituted cyclopropenones [104], but the intermediacy of a tricyclo[3.1.0.0^{2,4}]-hexane derivative cannot be fully excluded.

2.2.5 Seven-Membered Rings

Two approaches to seven-membered rings have been realized with the aid of transition metal catalysts:

1) [3+2+2]-cycloadditions (Eq. 44)
2) [2+2+2+1]-cycloadditions (Eq. 45)

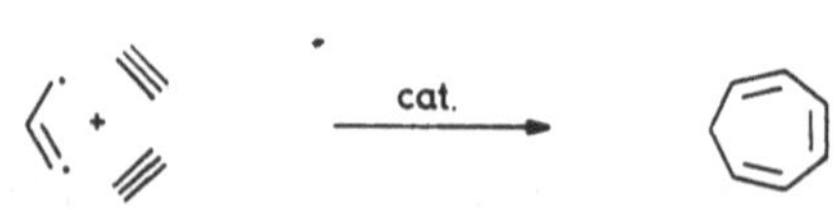

(44)

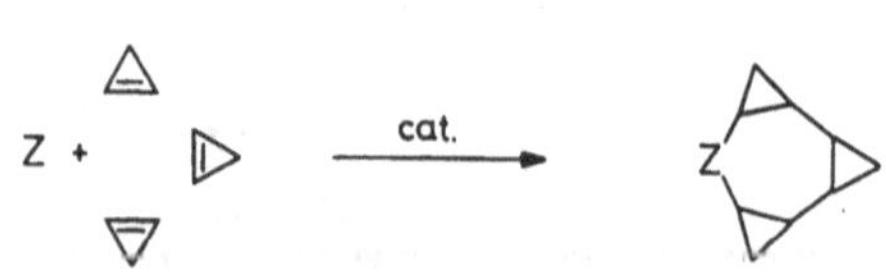

(45)

The first approach has already been mentioned in the preceeding chapter and will not be discussed further.

A [2+2+2+1]-cycloaddition could be achieved in the case of the Pd(0) catalyzed cotetramerization of 3,3-dimethylcyclopropene with carbon monoxide to give the hexamethyl-σ-trishomotropone *36* in nearly quantitative yield [105] (Eq. 46).

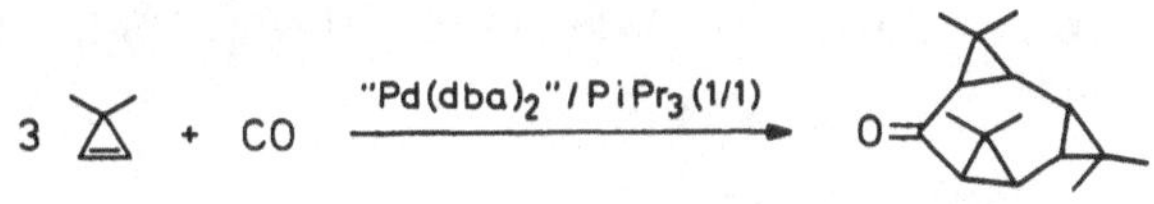

36 (46)

To our knowledge this remarkable reaction is the only example of a transition metal catalyzed [2+2+2+1]-cycloaddition that has been discovered so far [92].

2.2.6 Eight-Membered Rings

This paragraph has only been included for the sake of completeness. [2+2+2+2]-Cycloaddition products of cyclopropenes have only been obtained as side products of the [2+2]-cycloaddition reaction. In the presence of phosphane-free palladium(0) catalysts, 3,3-diorganylcyclopropenes undergo cyclotetramerization to give two isomers in low combined yields [73, 74] (Eq. 47).

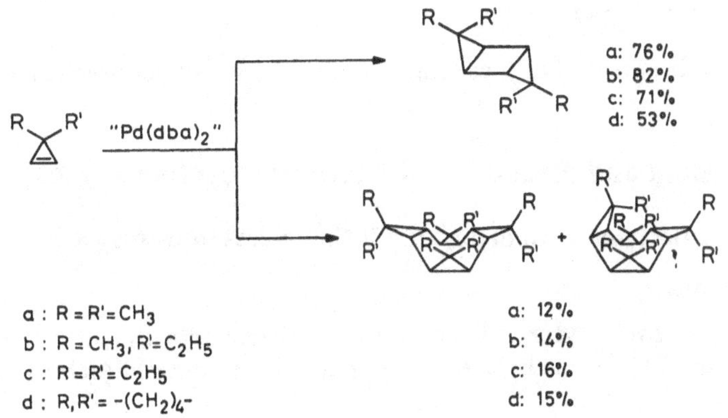

(47)

Thus the synthetic utility of this transformation is, at least at this time, somewhat limited.

2.3 Synthesis of Open-Chain Hydrocarbons

Cyclopropenes can be used in the preparation of conjugated trienes via formal recombination of two vinylcarbene units (Eq. 48).

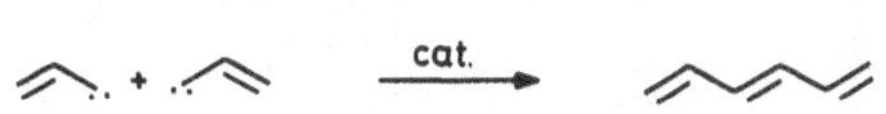

(48)

It has already been mentioned (see Chap. 2.2.1) that 1,3,3-trimethylcyclopropene is dimerized under the influence of Cu(I) or Cu(II) salts to give a 60% yield of 2,3,6,7-tetramethylocta-2,4,6-triene [64a]. In the case of tetramethylcyclopropene, the Cu(I) catalyst has to be activated by acrylonitrile [64b]. Otherwise the alcohol *37* is isolated (Eq. 49), indicating the intermediacy of a vinylcarbene species.

3-Cyclopropyl-3-methylcyclopropene and 3,3-dimethylcyclopropene undergo this reaction in CH_2Cl_2 even at temperatures below 0 °C [67a] (Eq. 50) in "high yields" (stereochemistry unknown).

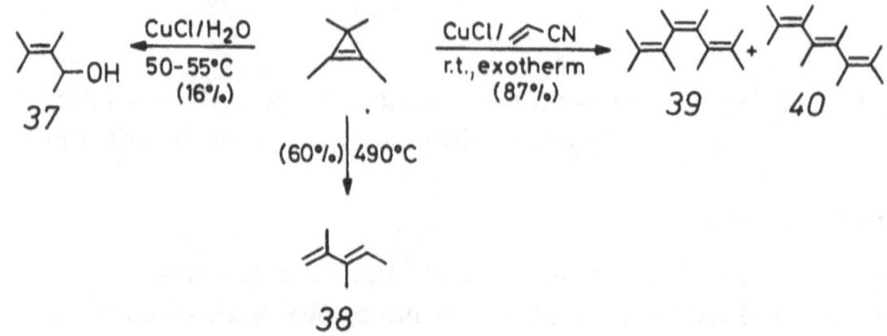

(49)

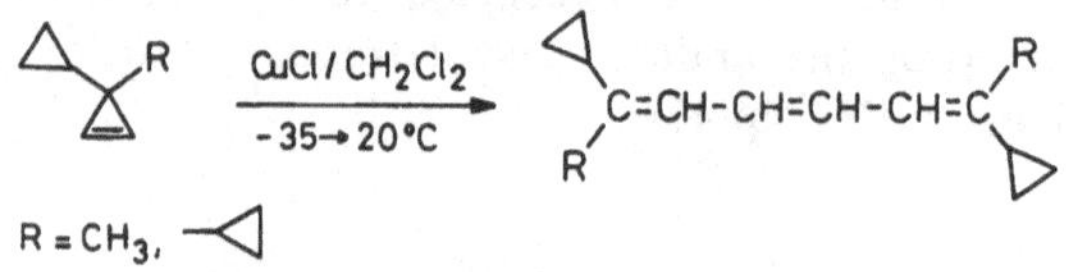

(50)

The synthetic potential of these transformation has not yet been explored.

3 Preparation and Reactions of Methylenecyclopropanes

3.1 Preparation and Properties of Methylenecyclopropanes

3.1.1 Preparation

During the past twenty years [106] a number of efficient and straightforward syntheses of the various types of methylenecyclopropanes have emerged (Scheme 2).

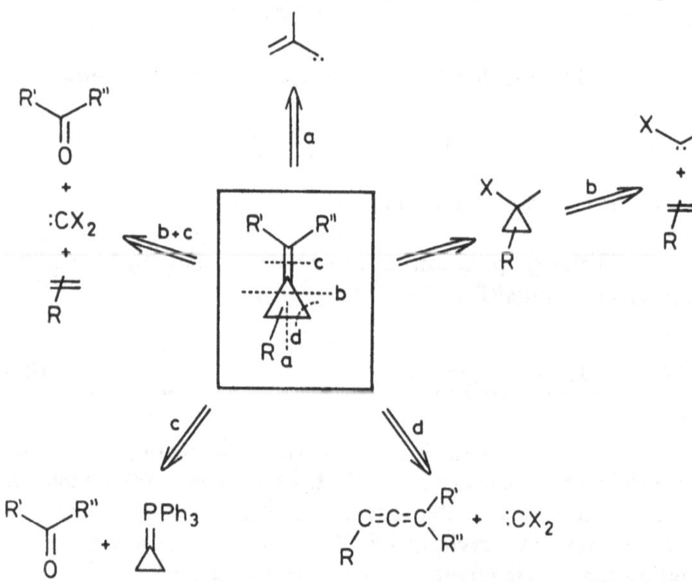

Scheme 2. Retrosynthetic analysis of methylene- and alkylidenecyclopropanes

These methods form the basis of the synthetic utilisation of methylenecyclopropanes on a wider scope.

Methylenecyclopropane itself is best prepared by base catalyzed rearrangement of methylcyclopropene which is formed by an intramolecular reaction of the vinylcarbene (Route a) generated from commercially available methallylchloride. The latter was reported to react with $NaNH_2$ in anhydrous THF to give pure 1-methylcyclopropene in 40% yield [107]. Appropriate variations of the base and/or the reaction conditions allow to prepare methylenecyclopropane in one or two steps, respectively, and in 70% overall yield [1-5] (Eq. 51). Following the procedure given in Ref. 5), methylenecyclopropane can be obtained even on a kilogram scale.

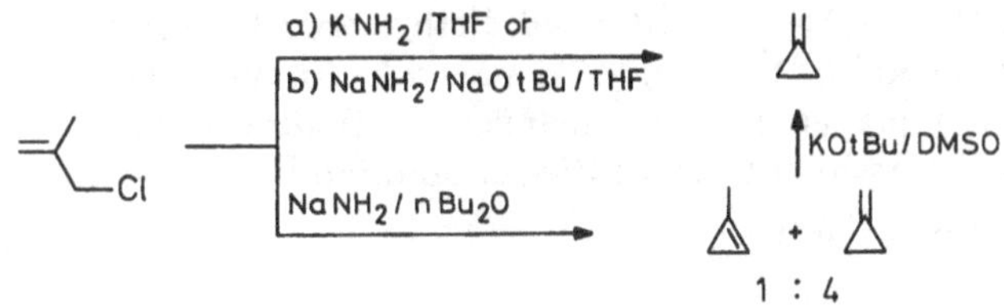

(51)

Extensions of this method seem to be limited to ethallylchloride from which ethylidenecyclopropane can be obtained in an overall yield of 47% [7].

The best method for the preparation of ring-alkylated and -arylated methylenecyclopropanes proves to be the dehydrochlorination of 1-chloro-1-methylcyclopropanes (Route b). The latter are easily obtained from alkenes and 1,1-dichloroethane in the presence of a suitable base [6] (Eq. 52).

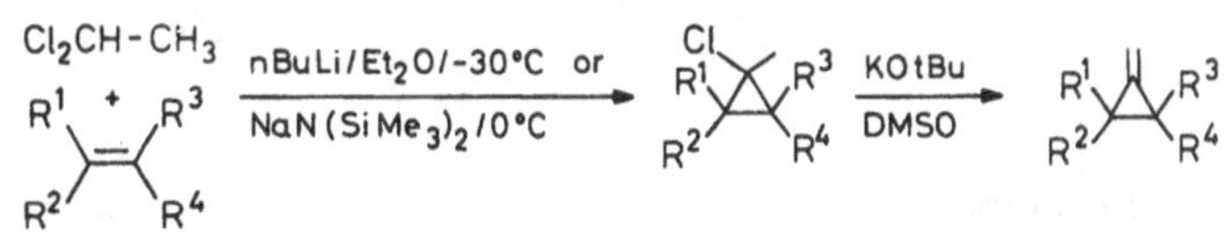

(52)

These reactions are limited to 1,1-dichloroethane as a carbene source, since higher homologues give poor yields (e.g. 15% in the case of 1,1-dichloropropane). All kinds of alkylated and arylated alkenes as well as conjugated and non-conjugated dienes may be used as carbenophiles. In all cases, the stereochemistry of the alkene is retained. Carbonyl groups attached to the alkene must be protected [108]. Starting with 1,1-dihalo- or 1-halocyclopropanes (halogen = Cl, Br), good yields of methylenecyclopropanes via dehydrohalogenation are only obtained, when the three-membered ring is alkylated in the 2- and 3-position [109-112] (Eq. 53–56).

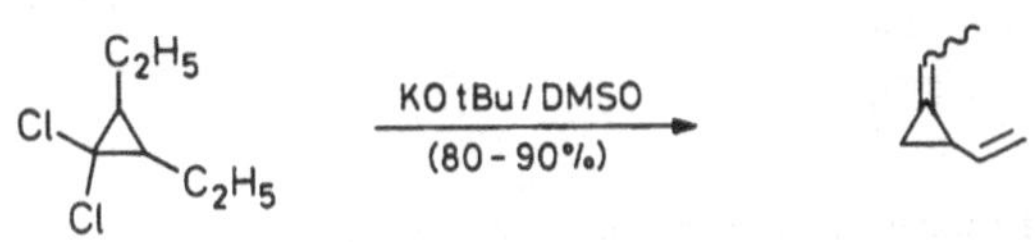

(53)

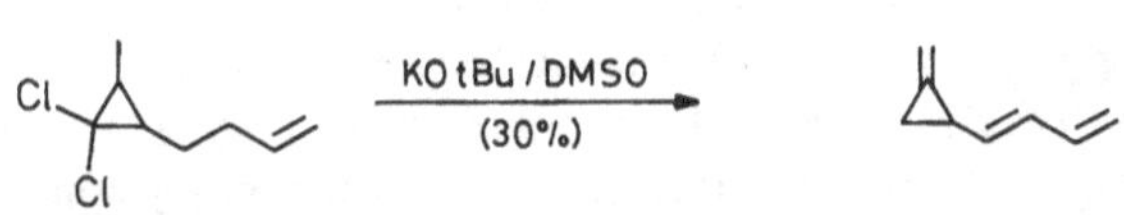

(54)

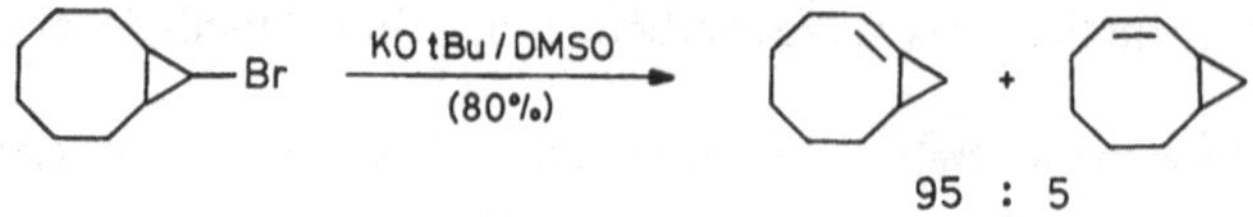

(55)

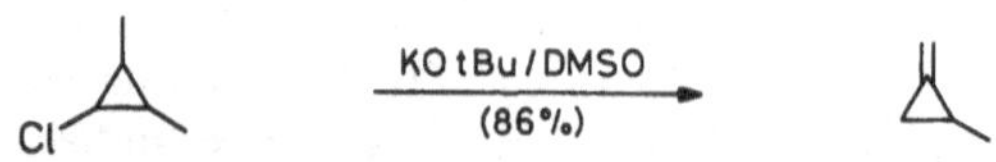

(56)

The starting materials are best prepared by the method of Makosza [113] (for dihalocyclopropanes) or via halocarbene addition to alkenes [114].

Another attractive method for the preparation of ring-substituted methylenecyclopropanes is the alkylation of lithiated methylenecyclopropane [115]. Until now, only trimethylsilylated [116, 117] (Eq. 57) and α-hydroxyalkylated [115] derivatives (Eq. 58) have been prepared in this manner. Methylenecyclopropanes with aryl groups at the vinylic position undergo Ziegler-addition [118] of *n*-BuLi under surprisingly mild conditions [119] (Eq. 59).

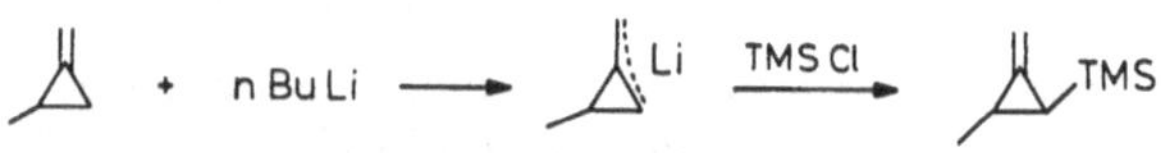

(57)

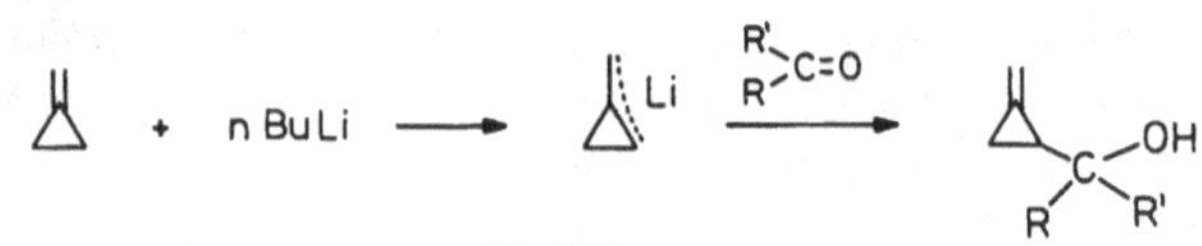

(58)

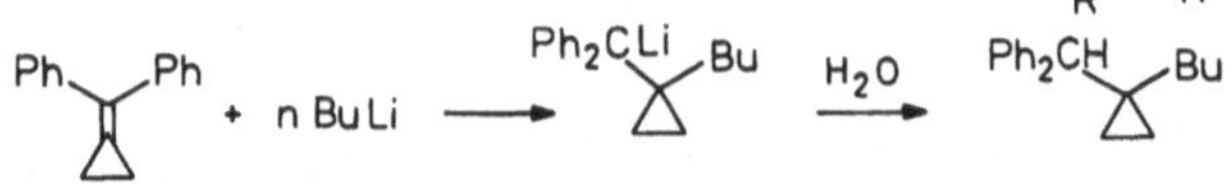

(59)

For a successful preparation of alkylidenecyclopropanes, two different strategies are available.

Wittig olefination of aldehydes and ketones with triphenylcyclopropylidenephosphorane (Route c) leads to a wide range of (organylmethylene)- or (diorganylmethylene)-cyclopropanes in satisfactory to good yields [120] (Eq. 60).

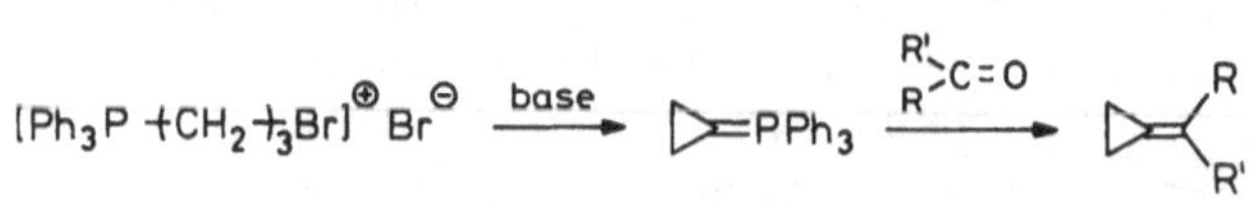

(60)

R = H′, R′ = Ph; R = R′ = Ph [120b]
R = Et, R′ = Bu [120b]
R, R′ = $-(CH_2)_5-$ [120c, d]
R = CH_3, R′ = $COCH_3$ [120f]

The olefination of cyclopropanone hemiacetales with alkylidenephosphoranes has also been published [121] but the yields are rather low.

The second strategy (Route b + c) appears to be more versatile since it also provides ring-substituted alkylidenecyclopropanes in an easy way. The key-step in this sequence is a Peterson olefination, starting with 1,1-dibromocyclopropanes [122a–c],

1,1-bis(phenylseleno)cyclopropanes [122c)] or 1,1-bis(phenylthio)cyclopropanes [122d)] (Eq. 61).

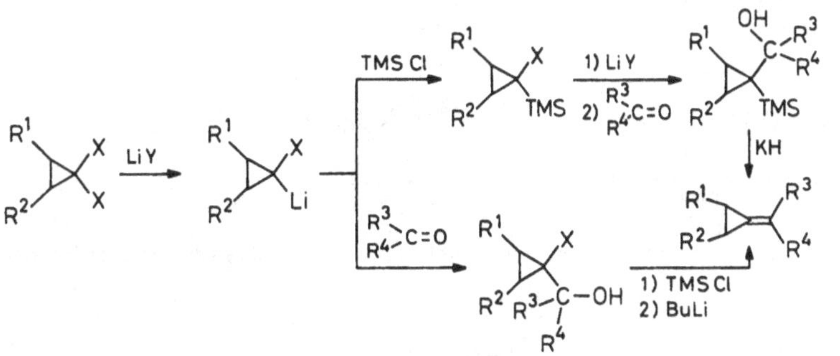

(61)

$R^1 \rightarrow R^4$ = H, Alkyl
X = Br [122a–c)], SePh [122c)], SPh [122d)]
Y = C_4H_9 [122a–c)], 1-(dimethylamino)naphtalenide [122d)]

Monoaddition of carbenes or carbenoids to allenes (Route d) has also been achieved, but addition of two carbenes to give spiropentane derivatives is difficult to avoid. Examples of successful applications of this approach are dihalocarbene additions to substituted allenes [123)] (Eq. 62) to give dihalomethylenecyclopropanes in good yields.

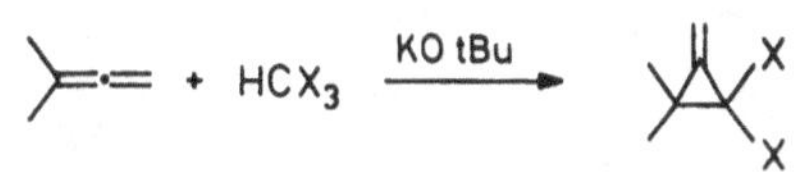

(62)

X = Cl, Br

Copper-catalyzed [124a–c)] or light-induced [124c–d)] decompositions of diazoalkanes in the presence of allenes result in the formation of spiropentanes as the main products. Some reports have been published concerning the addition of unsaturated carbenes to alkenes to give methylenecyclopropanes (Route b) [125)]. Unfortunately, this method seems not to be practicable for preparations on a somewhat larger scale.

3.1.2 Properties

Methylenecyclopropanes are highly strained molecules [37)]. Nevertheless, they are perfectly stable at ambient temperature and even some naturally occurring compounds (e.g. hypoglycin A [126)]) contains a methylenecyclopropane unit. They have therefore attracted considerable interest both in mechanistic [127)] as well as synthetic studies [128)].

The molecular structure of methylenecyclopropane has been determined by microwave spectroscopy [129)] and may be compared with that of 1-(diphenylmethylene)-cyclopropane, which has been established by a three-dimensional X-ray analysis [130)] (Fig. 5).

The exocyclic double bond imposes steric strain on the three-membered ring which

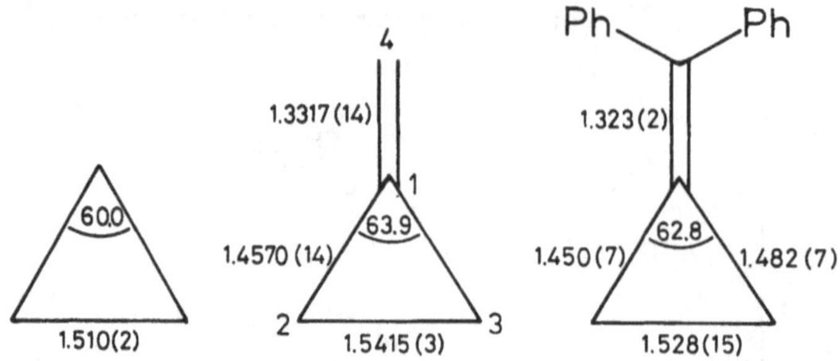

Fig. 5. Structural data for cyclopropane [34], methylenecyclopropane [129] and 1-(diphenylmethylene)-cyclopropane [130]

is reflected by a lengthening of the C(2)–C(3) bond (compared to cyclopropane) to 1.5415(3) A, and an increase of the C(2)–C(1)–C(3) angle to 63.9°.

The spectroscopic data of methylenecyclopropane and some of its derivatives have been published: NMR [131], IR [132], Raman [132c], microwave [129], UV [133] and PE [134].

It is now commonly accepted that most carbocycles are strained to some degree. One might anticipate that the absolute strain energies calculated for a given set of related molecules are directly proportional to their reactivities. Unfortunately, the

Table 3. Strain reliefed upon hydrogenation of double bonds in various strained hydrocarbons (kcal/mol) [33]

	−0.9		13.0
	−0.3		25.8
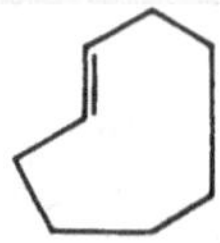	6		26
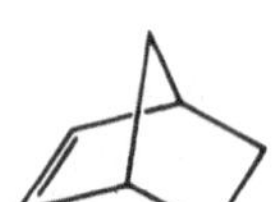	9.6		26

situation is much more complex. As an example one may consider the three smallest propellanes, [2.2.1]-, [2.1.1]- and [1.1.1]-propellane [135]. All three compounds have about the same strain energies but very different thermal stabilities and reactivities. This is due to a different degree of relief of strain upon homolytic cleavage of the "strained" C—C bond. Thus the strain energies of the intermediates (in the above case the corresponding diradicals) are of equal importance [136]. Especially in transition metal catalyzed reactions this point is of great importance. The most reliable source of information about the relative reactivities of unsaturated (olefinic) carbocycles is probably the strain relieved upon hydrogenation of the double bond (Table 3).

According to the data in Table 3 the reactivity of methylenecyclopropanes should lie somewhere between that of norbornene and a cyclopropene. In practice, the situation is even more complex (and sometimes puzzling) because there are three different "strained" bonds within a methylenecyclopropane: C(1)–C(4), C(2)–(3) and C(1)–C(3). Moreover, little is known about the strain energies of the organometallic intermediates involved in transition metal catalyzed reactions of methylenecyclopropanes.

Methylenecyclopropanes readily rearrange by cleavage of the C(2)–C(3) bond, propably via a perpendicular singlet trimethylenemethane diradical (TMM) (Eq. 63) [137,138].

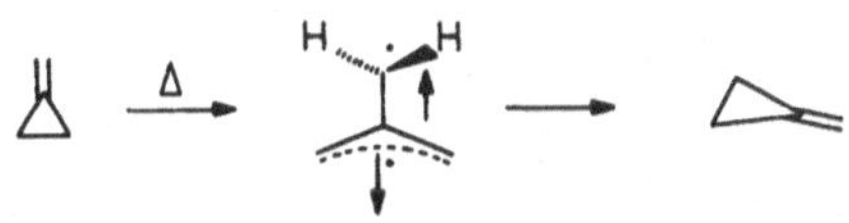

(63)

This perpendicular singlet TMM is calculated to be 8.4 to 25 kJ/mol more stable than the planar singlet TMM [139]. Upon γ-irradiation the TMM diradical can be generated even at −196 °C and observed spectroscopically [140] (Eq. 64).

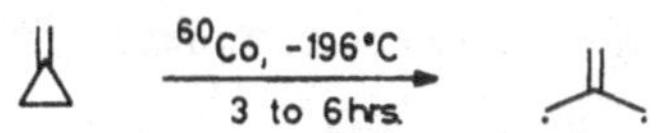

(64)

Methylenecyclopropanes also undergo reactions characteristic for reactive olefines, such as electrophilic additions, radical additions, additions of carbenes and nitrenes as well as the various other types of cycloadditions (Scheme 3).

The latter are of special synthetic interest since they result in the formation of two new C—X bonds (X = C, N). As shown in Scheme 3, thermally induced cycloadditions can only be achieved at high temperatures or when the methylenecyclopropane or the cosubstrate are activated by strong electron-withdrawing groups, such as halogen or CN. As will be outlined in the following chapters, a number of cycloaddition reactions of methylenecyclopropanes can be achieved under moderate conditions and in satisfactory to good yields in the presence of suitable transition metal catalysts.

But before we are discussing the results obtained so far in this area, we will briefly summarize the types of stoichiometric reactions that methylenecyclopropanes undergo with transition metal complexes. The following reactions have been observed (Scheme 4):

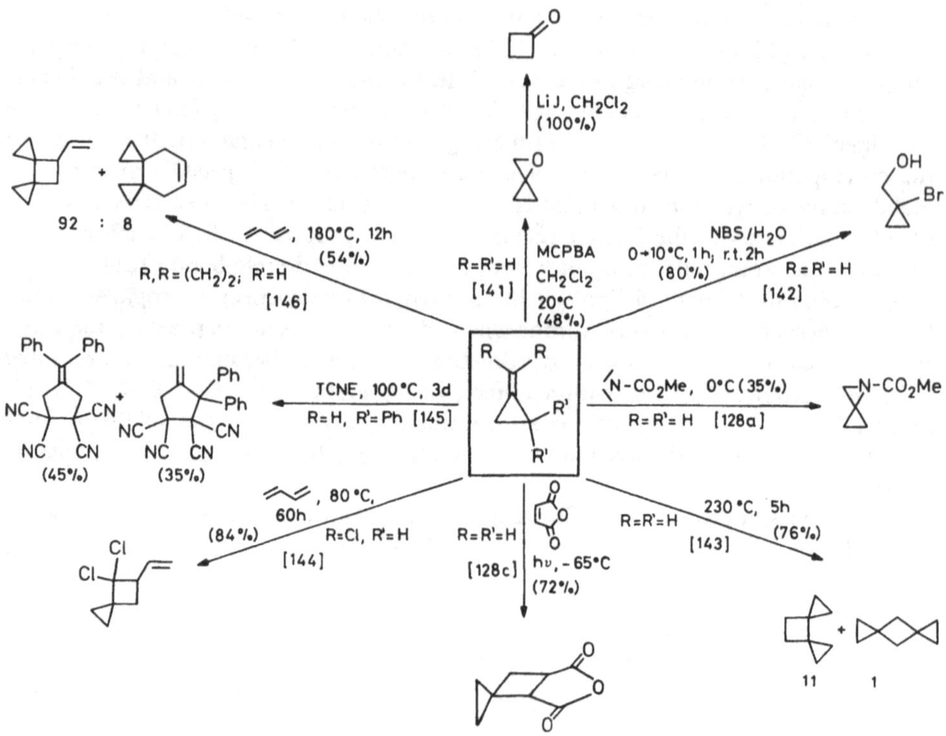

Scheme 3. Selected organic reactions of methylenecyclopropanes

1) π-Complexation of one [147–149] or two [156] methylenecyclopropanes leading to complexes of Type I or II.
2) Ring-opening between C(1) and C(2) to give 1,3-diene complexes [150,151], mostly via I.
3) Ring-opening between C(2) and C(3) to give trimethylenemethane complexes III [153,154]. In some cases the intermediacy of I could be proven [153].
4) Chloropalladation to give allylcomplexes of Type IV [155].
5) Oxidative coupling to give metallacyclopentanes V [157,158]. Here the π-complexes II should be the precursors, which has been confirmed in some cases [157].

For catalytic transformations of methylenecyclopropanes so far nickel(0) and palladium(0) complexes have been used successfully. From the metal complexes mentioned in Scheme 4, the Types I, II, III, and V play important roles as intermediates in cyclo- and cooligomerization reactions whereas the Sequence I → VI is responsible for isomerizations.

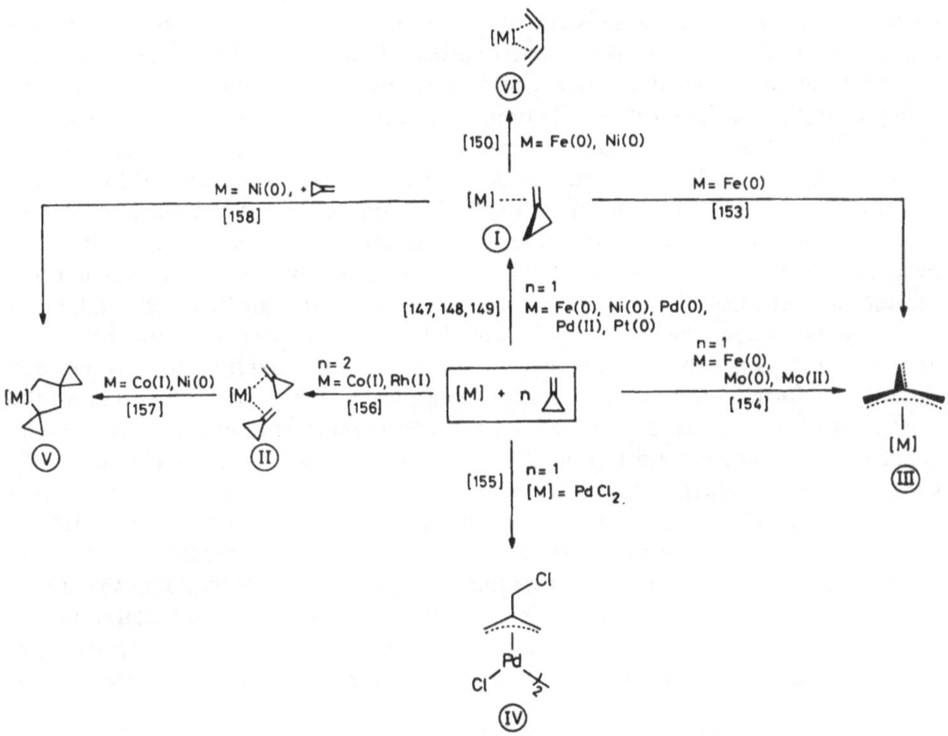

Scheme 4. Reaction types of methylenecyclopropane with transition metal complexes

3.2 Synthesis of Four-Membered Rings

Among the numerous ways to prepare C_4-rings [159] the [2+2]-cycloaddition of two multiple bonds appears to be the most direct approach and potentially the most powerful one. Corey's caryophyllene synthesis may serve as a prominent example [160] (Eq. 65).

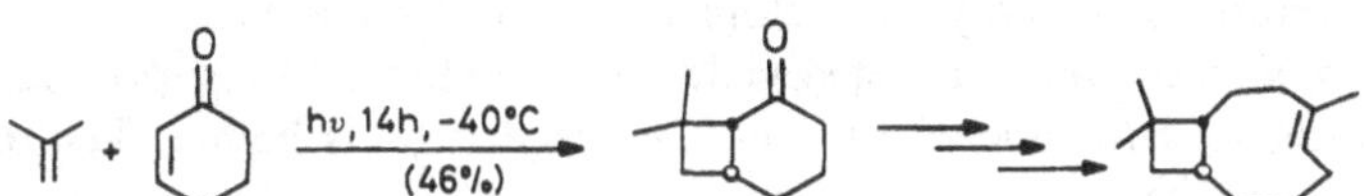

(65)

There are, however, some severe drawbacks associated with [2+2]-cycloadditions; although an in depth discussion of these is beyond the scope of this article, the mayor ones will be mentioned briefly. Thermal [2+2]-cycloadditions are essentially limited to a few cases. This is due to the fact that the cycloaddition of two alkenes in a $[\pi^2s + \pi^2s]$ manner violates the rule of conservation of orbital symmetry [161]. On the other hand, a $[\pi^2s + \pi^2a]$ mechanism, though symmetry allowed, is unlikely because of the twisting energy necessary to obtain overlapping in the transition state (ketenes and other linear molecules can react by this mechanism because the twisting energy can be minimized). Thus these reactions are usually carried out at high temperatures (100 to 230 °C) and under pressure, reflecting a diradical (or dipolar) mechanism. Some of the drawbacks of this approach can be overcome by photochemical excitation. In this case a greater variety of alkenes (though by no means all) can be brought to reaction under mild conditions. Unfortunately, a clean and high-yielding photochemical [2+2]-codimerization is the exception rather than the rule [162]. In the late sixties, Mango introduced the postulate that [2+2]-cycloadditions should be symmetry allowed in the presence of transition metal catalysts ("forbidden-to-allowed catalysis")' For instance, the Ni(0) catalyzed cyclodimerization of 1,3-butadiene to *cis*-1,2-divinylcyclobutane was considered to be a suprafacial concerted process [163b]. Shortly thereafter, experimental evidence was obtained which ruled out such a mechanism [164]. Further studies revealed that all transition metal catalyzed [2+2]-cycloadditions, yielding cyclobutane derivatives, proceed stepwise via metallacycloalkanes as intermediates [165, 166, 167]. Recently, Hoffmann et al. have pointed out that the formation of these intermediates as well as their transformation into the products can be considered themselves as concerted processes as long as characteristic geometric requirements are fulfilled [168, 169]. The transition metal catalyzed [2+2]-cycloadditions of alkenes are — at least at the moment — restricted to strained molecules such as cyclopropenes, methylenecyclopropanes or norbornene [92]. In this context, methylenecyclopropanes are of special interest from a synthetic point of view. For instance, the reader can easily recognize the hidden methylenecyclopropane unit in caryophyllene. Therefore, systematic model studies on the transition metal catalyzed cycloaddition reactions of methylenecyclopropanes were undertaken, especially in our laboratories in Mülheim. The results of these studies are summarized in the following.

In the presence of Ni(0) catalysts, methylenecyclopropanes cyclodimerize at temperature as low as —15 °C. The chemoselectivity of these reactions strongly depends on the substitution pattern of the substrates. [2+2]-Cyclodimerizations are restricted to methylenecyclopropanes bearing no further substituents at the exocyclic double bond. The substitution pattern at the threemembered ring (R = alkyl, aryl) determines whether four-membered rings, five-membered ring or open-chain products (or a mixture of all those) are obtained.

$Ni(cod)_2$ or mixtures of $Ni(cod)_2$ with an electron deficient alkenes (e.g. dialkyl fumarate or maleic anhydride) have been found to be the most efficient catalysts for the cyclodimerization of methylenecyclopropane and 2-methylmethylenecyclopropane [170, 171]. With $Ni(cod)_2$ the combined yields of cyclodimerization products are lower, but the ratio or four-membered to five-membered rings is higher. The reverse holds for the modified catalysts (Eq. 66).

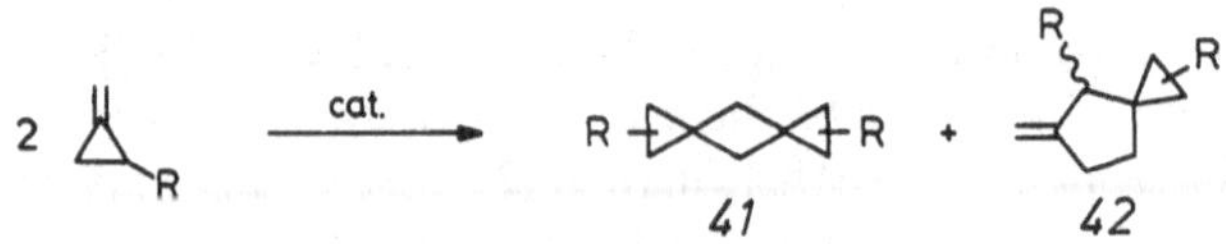

41 42 (66)

cat.	R	yield (*41* + *42*)	ratio *41*:*42*
$Ni(cod)_2$	H	50%	20:80
$Ni(cod)_2$	CH_3	25%	40:60
$Ni(cod)_2$/dialkyl fumarate	H	80%	5:95
$Ni(cod)_2$/dialkyl fumarate	CH_3	85%	11:89

For the transformation of more highly substituted methylenecyclopropanes, trialkylphosphane modified Ni(0) catalysts are more effective [172]. Representative examples are the cyclodimerizations of 2,2-dimethyl- and 2,2,3,3-tetramethylmethylenecyclopropane (Eqs. 67 and 68).

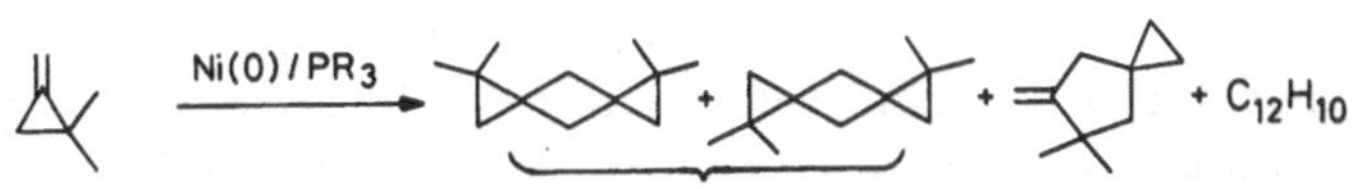

43 44 45 (67)

R_3P	yield (*43* + *44* + *45*)	ratio *43*:*44*:*45*
PEt_3	92%	29:40:31
$P(iPr)(tBu)_2$	86%	76: 5:19

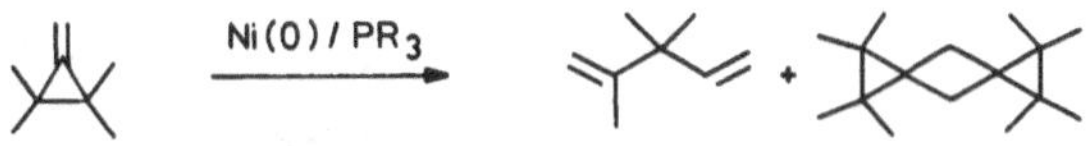

46 47 (68)

R_3P	yield (*46*)	yield (*47*)
PEt_3	68%	18%
$P(iPr)(tBu)_2$	82%	—
$P(C_6H_{11})_3$	24%	35%

While 2,2-dimethylmethylenecyclopropane can be cyclodimerized in 80 to 90% yield, tetramethylmethylenecyclopropane gives mainly the ring-opened product 2,3,3-trimethyl-1,4-pentadiene [172]. Control of stereochemistry in these reactions is low. However, the observed regiospecifity of four-membered ring formation is surprising (Eq. 69).

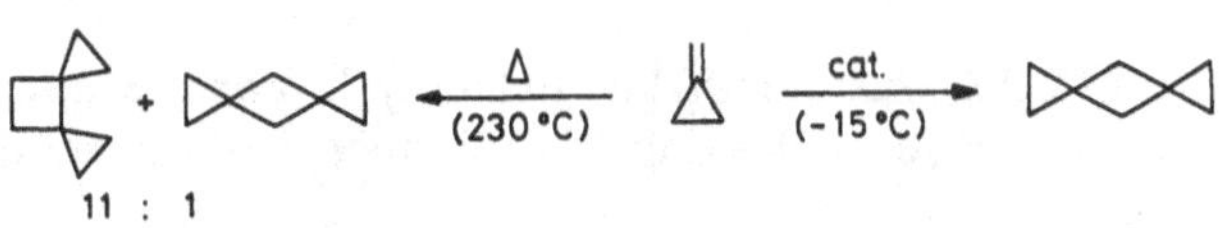

(69)

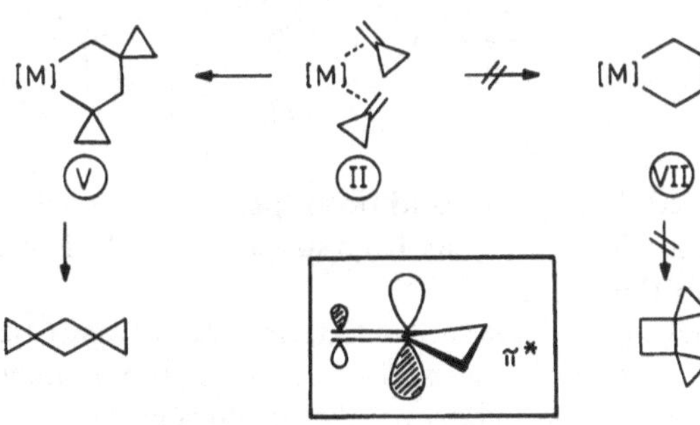

(70)

Dispiro[2.1.2.1]octane is obtained as the only product in the presence of Ni(0), in contrast to the result of the thermal cyclodimerization [143]. According to the polarization of the π^*-orbital of methylenecyclopropane one would expect the symmetrical nickelacyclopentane VII as an intermediate [168], leading to dispiro[2.0.2.2]octane (Eq. 70). Obviously, steric constraints in the nickelorganic intermediate [148b] are responsible for the exclusive formation of dispiro[2.1.2.1]octane.

Metal-catalyzed [2+2]-codimerizations of methylenecyclopropanes with other alkenes are limited to a few cases. Again the formation of the formal [3+2]-cycloadducts is competing. As cosubstrates, strained alkenes (Eqs. 71 and 72) and alkyl acrylates could be applied successfully, the latter being of more interest from a synthetic point of view.

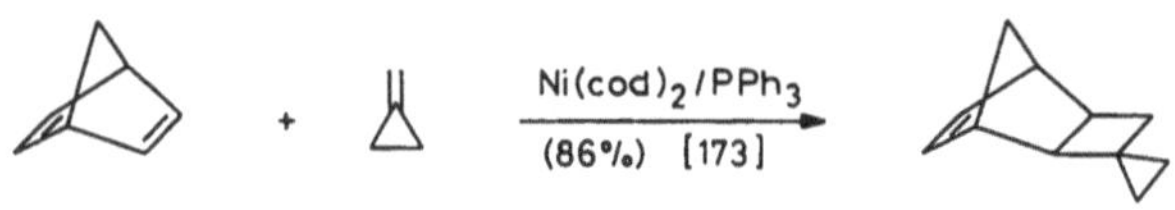

(71)

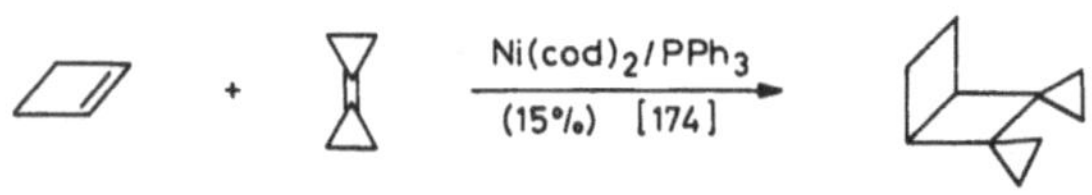

(72)

In the alkyl acrylate reactions the substitution pattern of the methylenecyclopropanes clearly controls the selectivity. While with methylenecyclopropane and its 2-methyl derivatives only five-membered rings are obtained, 2,2-dimethylmethylenecyclopropane gave four-membered rings in moderate yields [175] (Eq. 73).

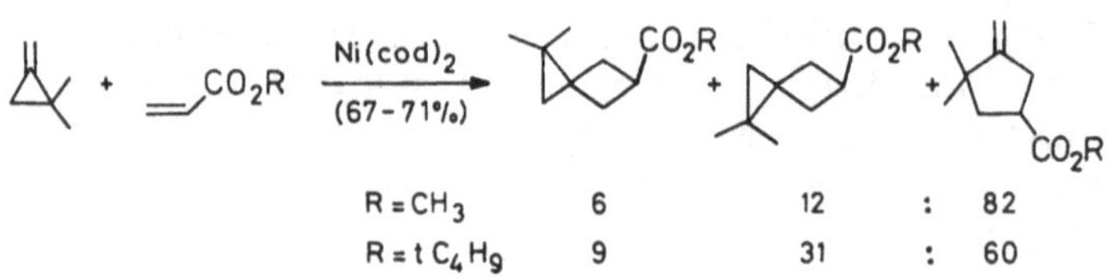

(73)

Finally, in the case of 2,2,3,3-tetramethylmethylenecyclopropane the corresponding cyclobutane product is formed as the only cycloadduct, besides small amounts of isomerization product [175] (Eq. 74).

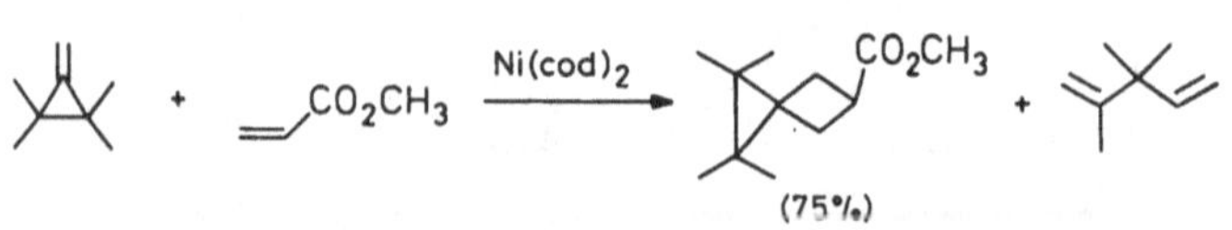

(74)

All these Ni(0) catalyzed [2+2]-cycloadditions proceed through nickelacyclopentane derivatives as intermediates. This has been supported by the isolation of model complexes and the study of their reactivity [148b, 156b].

The results of these model studies have shown that transition metal catalyzed [2+2]-cycloadditions of methylenecyclopropanes are — at least at the moment — not very useful for the controlled formation of four-membered carbocycles.

3.3 Synthesis of Five-Membered Rings

3.3.1 Methylenecyclopentanes

3.3.1.1 Introduction

Retrosynthetic analysis of six-membered ringformation almost always boils down to a Diels-Alder reaction [176] or a Robinson annelation [177] (or variations thereof) as the crucial C—C bond forming step. Both methods have in common that more than one carbon—carbon bond is formed in a one pot reaction which allows a rapid and efficient construction of complex organic molecules from rather simple building blocks. No such general tool exists for the formation of carbocyclic five-membered rings.

Since the early sixties an impressive array of natural products containing five-membered rings and possessing interesting biological activities has been uncovered. This initiated a search for new methodologies for their preparation [178]. The iridoid glycoside loganin (*48*) may serve as a prominent example.

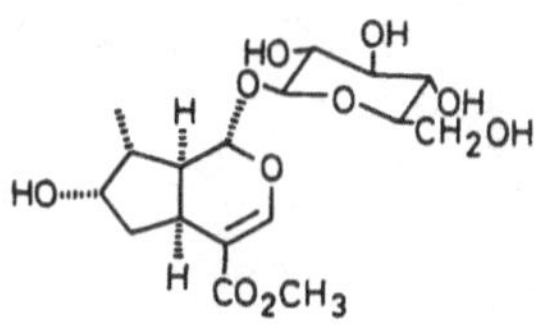

48

(74a)

Loganin turned out to be the biological precursor of secologanin which itself is the biogenetic key compound to more than a thousand alkaloids and other natural products [179]. Moreover, non-natural products such as dodecahedrane [180] have been imagined and subsequently synthesized. As a result of these efforts, the past two decades have seen numerous multistep C_5 annelation procedures, mainly focussing on the intermediacy of 1,4-dicarbonyl compounds or their functional group equivalents [178]. The most straightforward strategy, a cycloaddition approach comparable to the Diels-Alder reaction (both schematically represented in Eqs. 75 and 76) has only recently been addressed.

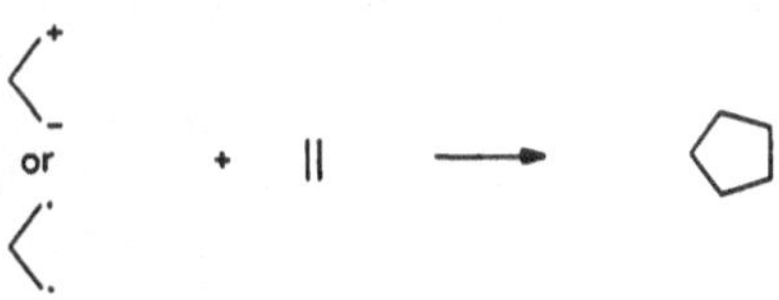

(75)

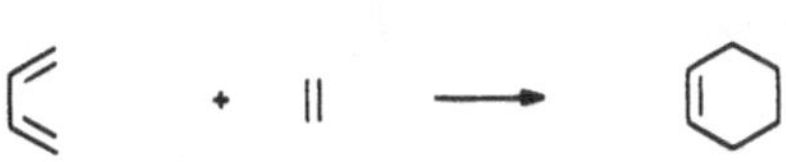

(76)

To realize a new strategy like the one outlined in Eq. 75, it is necessary to find a reasonable way to create the odd-numbered carbon fragment. Faced with such a problem the view almost automatically turns to the trimethylenemethane (TMM) diradical (cf. Eq. 64), a species well studied from a physical-organic perspective [181]. It provides not only the necessary three-carbon fragment but also an exocyclic methylene group for further structural elaboration. The lack of a suitable method to generate this synthon on a preparative scale prevented its application for years [182]. Recently, this problem has been solved by the discovery of suitable TMM-precursors that can be transformed into the TMM-synthon via transition metal catalysis. One class of precursors consists of conjunctive reagents, exemplified by compounds *49* [183] and *50* [184]. In the presence of palladium(0) catalysts, zwitterionic TMM-palladium complexes are formed which undergo a Michael type reaction with electron-deficient alkenes followed by ring-closure to the corresponding methylenecyclopentane derivatives [183c] (Eq. 77).

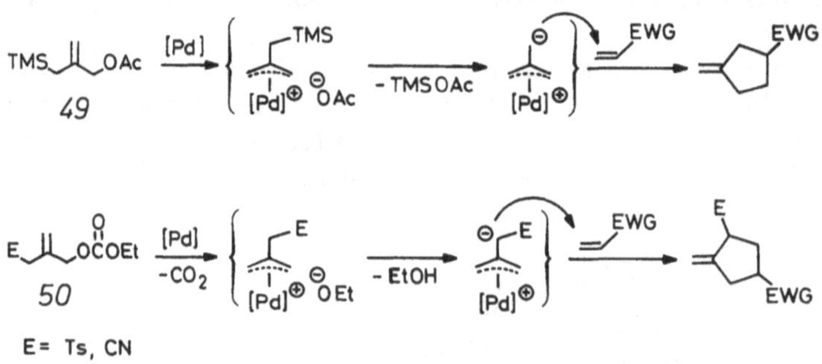

(77)

Some years ago, we began to explore the chemistry of another class of precursors — methylenecyclopropanes — in the presence of transition metal catalysts. Since many of these compounds are now easily available on a large scale, it occurred to us that they might be excellent starting materials to serve the purposes of a [3+2]-cycloaddition approach. In contrast to the above mentioned conjunctive reagents, two different types of synthons may be generated from methylenecyclopropanes that are both suitable for [3+2]-cycloadditions. This is outlined in a purely formalistic manner in Scheme 5.

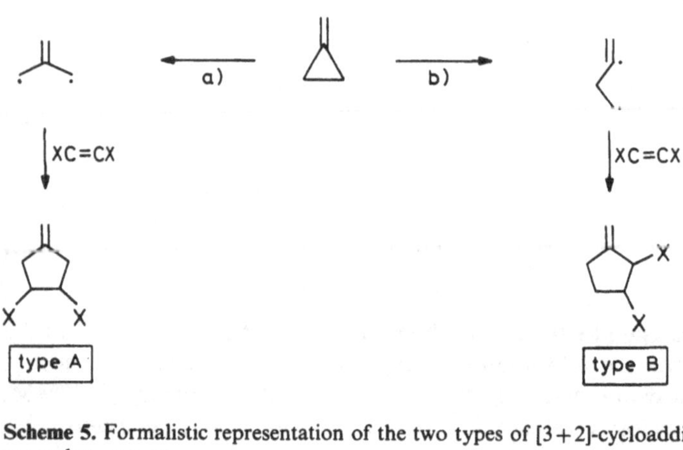

Scheme 5. Formalistic representation of the two types of [3+2]-cycloadditions observed for methylenecyclopropanes

Indeed, both kinds of cycloaddition products (Type A and Type B) can be obtained in the presence of Ni(0) catalysts while Pd(0) catalysts exclusively lead to Type A codimers. The real course of these reactions however is somewhat more complicated. While distal ring-opening via Route a really leads to cycloaddition products of Type A, proximal ring-opening via Route b results only in an isomerization of methylenecyclopropane. Cycloaddition products of Type B are obtained indirectly via oxidative coupling of two alkene units with low-valent nickel followed by a cyclopropylmethyl/3-butenyl rearrangement [22, 148b]. Reductive elimination terminates the catalytic cycle (Eq. 78).

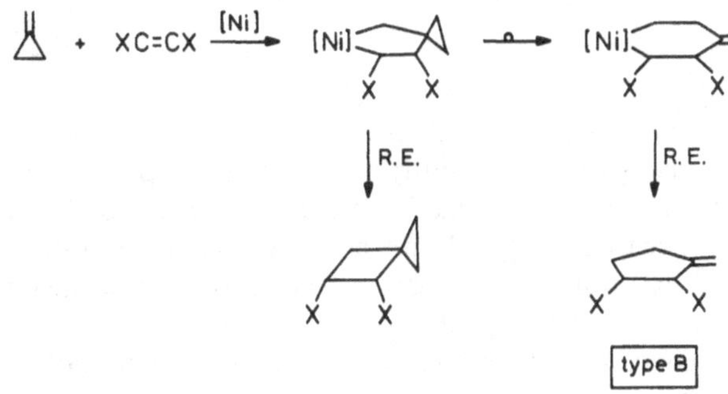

(78)

During our investigation of these codimerizations it turned out that with Ni(0) catalysts not only the metal determines the course of the reaction, but also several other factors; e.g. the kind and number of the ligands bonded to the nickel, the kind, number and position of substituents on the methylenecyclopropane and the electronic properties of the second olefine. These observations make it nearly impossible to predict, whether methylenecyclopentanes of Type A or B will be the products.

The cyclodimerization of methylenecyclopropanes has already been discussed in Sect. 3.2. For the sake of completeness we just mention a some what misleading reaction, in which methylenecyclopropane apparently serves only as a source of butadiene (Eq. 79) [185] before we turn our attention to the more interesting codimerization reactions.

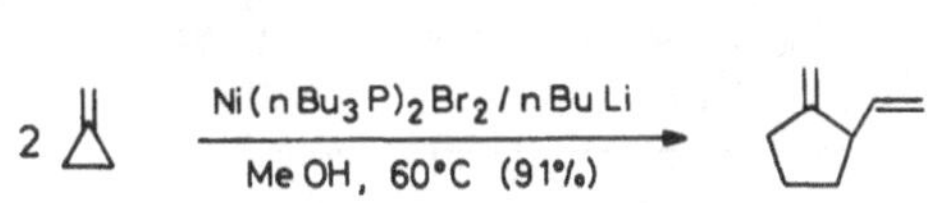

(79)

In the following we will first discuss the metal catalyzed codimerizations of methylenecyclopropane before we focus our attention to the behaviour of differently substituted methylenecyclopentanes.

3.3.1.2 Codimerizations with Methylenecyclopropane

In the presence of "naked nickel", methylenecyclopropane can be codimerized with alkyl acrylates, alkyl crotonates and alkyl maleates [175, 186, 187] giving Type B cycloadducts in moderate to excellent yields (Eq. 80).

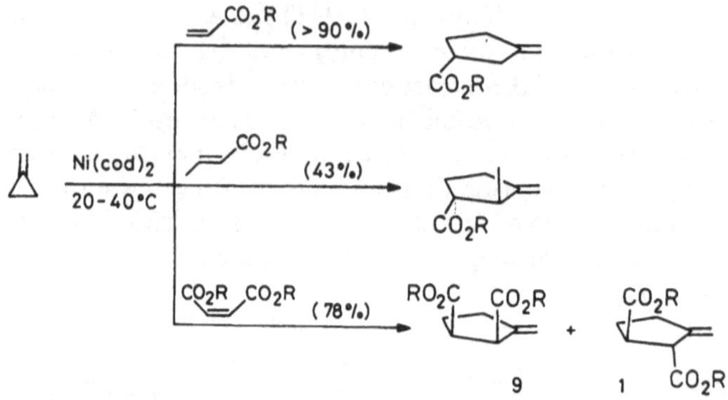

(80)

These reactions proceed smoothly at 20 to 40 °C with high regio- and stereoselectivity. When alkyl acrylates with an optically active alkyl group are applied, 3-methylenecyclopentyl carboxylates are obtained with up to 64% d.e. [188]. Interestingly, dialkyl fumarates and maleic anhydride do not react as cosubstrates but function as controlling ligands resulting in a considerable enhancement of the cyclodimerization of methylenecyclopropane (see p. 106).

The use of triphenylphosphane- or triarylphosphite-modified Ni(0) catalysts (e.g. $Ni(cod)_2/PPh_3$) demands higher reaction temperatures and leads to a decrease in stereoselectivity [189]; e.g. pure *trans*-alkenes give stereoisomeric mixtures of methylenecyclopentanes (Eq. 81). On the other hand more electron deficient olefines, e.g. crotonaldehyde, also codimerize in the presence of these catalysts.

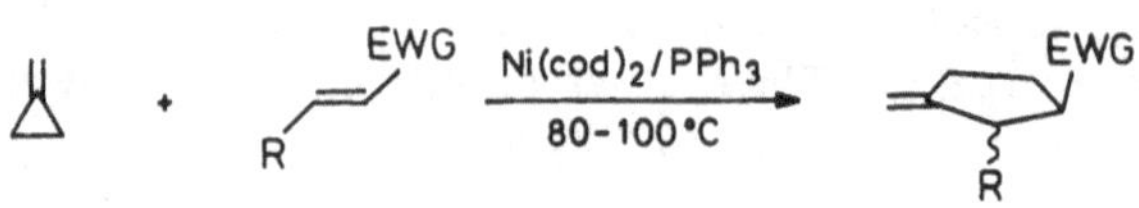

(81)

R	EWG	yield	*cis/trans*
H	CO_2CH_3	55%	—
CH_3	CO_2CH_3	55%	22:78
CH_3	CHO	54%	9:91
n-C_3H_7	CO_2CH_3	50%	27:73
$CH_2CH_2CO_2CH_3$	CO_2CH_3	49%	28:72

Most interestingly, dialkyl maleates no longer give rise to Type B cycloadducts but end up exclusively as Type A cycloadducts [189] (cf. Scheme 5). The same has been found with electron deficient cosubstrates such as dialkyl fumarates or 2,3-di(methoxycarbonyl)norbornene [189] (Eq. 82).

The last few examples clearly demonstrate that these Ni(0) catalyzed reactions are extremly sensitive to changes in the electronic properties of the catalyst (or catalyst precursor) as well as the substrates.

In contrast to Ni(0), Pd(0) leads exclusively to distal ring-opening of methylenecyclopropane. Thus in the presence of an appropiate diylophile, Type A cycloaddition is observed, according to Eq. 83:

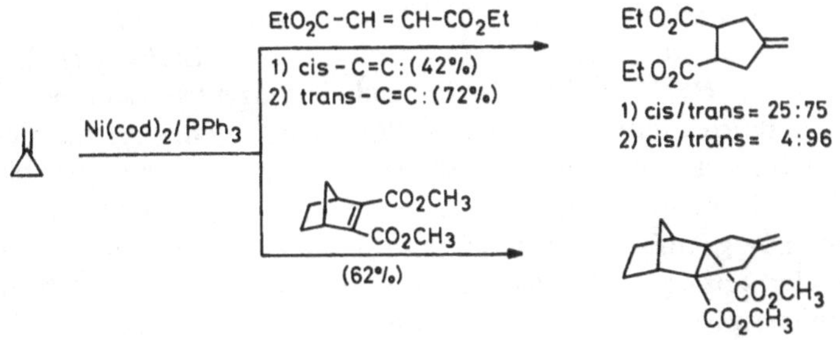

(82)

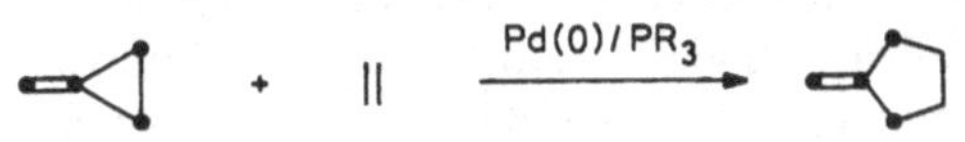

(83)

This unique behavior of Pd(0) catalysts allowed the development of a general method for the preparation of methylenecyclopentanes with a predictable substitution pattern.

For the choice of a suitable catalyst (or catalyst precursor) one has several options. The best results are usually obtained with a 1:1 mixture of Pd(0) and a phosphane or phosphite. As a source of Pd(0), thermally stable Pd(0) complexes such

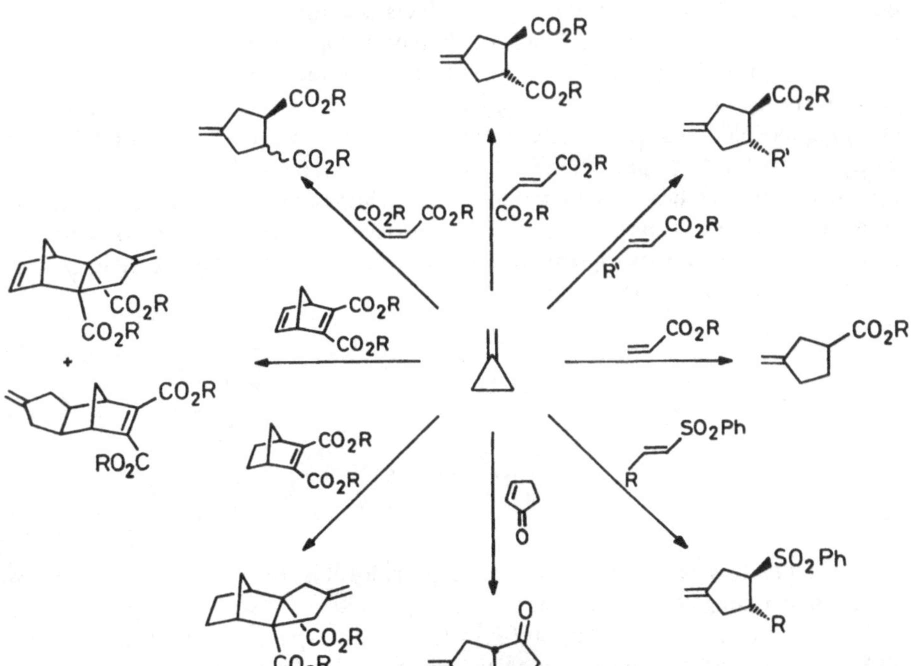

Scheme 6. Survey of Pd(0) catalyzed codimerizations of methylenecyclopropane with electron deficient alkenes

as bis(dibenzylideneacetone)palladium [190] can be used as well as in situ generated Pd(0) via reduction of Pd(II) salts (e.g. $Pd(acac)_2$ + Et_2AlOEt). In the case of $Pd(\eta^3\text{-}C_3H_5)(\eta^5\text{-}C_5H_5)$ [191,192] no further reducing agent is required. Tri-sec-alkyl-phosphane-modified catalysts (e.g. $Pd(0)/P(iPr)_3$) usually display the highest activity.

As cosubstrates a wide range of electron deficient alkenes can be brought to reaction [193] (Scheme 6). Using the above mentioned Ni(0) catalysts some of these codimerizations e.g. with 2-pentenone or with vinylsulfones, cannot be achieved and other cycloadditions e.g. with crotonates lead to different methylenecyclopentanes.

These codimerization reactions are mainly limited by the degree of π-bond strength of the electron deficient alkenes to Pd(0). Strongly bonded ligands may prevent any interaction of the metal with the methylenecyclopropane. Typical examples of too strongly bonded alkenes are maleic anhydride, acrolein and acrylonitrile. On the other hand, too weak interactions may result in cyclodimerization of the methylenecyclopropane rather than codimerization.

The following alkenes have been found to give high yields of codimerization products with methylenecyclopropane: acrylates, fumarates, maleates [193], cyclopenten-3-one [27], 2,3-dimethoxycarbonylnorbornene, 2,3-dimethoxycarbonylnorbornadiene [27] and some α,β-unsaturated sulfones [194]. The reactions proceed in a temperature range of 100 to 140 °C. It has been found that pumping a solution of educts into a preheated solution of the catalyst increases the yield [195].

Some of the reactions summarized in Scheme 6 deserve some further comments:

1. With dialkyl maleates isomerization at the metal takes place before codimerization. Thus a *cis/trans* mixture of cycloadducts is obtained; e.g. heating of a 1:1 mixture of dimethyl maleate and methylenecyclopropane in the presence of 0.2 mol% $(\eta^3\text{-}C_3H_7)(\eta^5\text{-}C_5H_5)Pd/P(iPr)_3$ for 8 hours to 100 °C gives a 77% yield of the cycloadduct as a 38:62 *cis/trans* mixture. Using the above mentioned pumping technique the yield of the cycloadduct is nearly quantitative after 2 hours at 130 °C with a 90% predominance of the *cis*-isomer [193,195].
2. Codimerizations with 2-cycloalkenones can be achieved only with 2-cyclopentenone in satisfactory yields (up to 78%) [195]. The Pd(0) catalyzed codimerization between 2-cyclohexenones and methylenecyclopropane has been reported to yield mono- and dialkylation products [196] (Eq. 84).

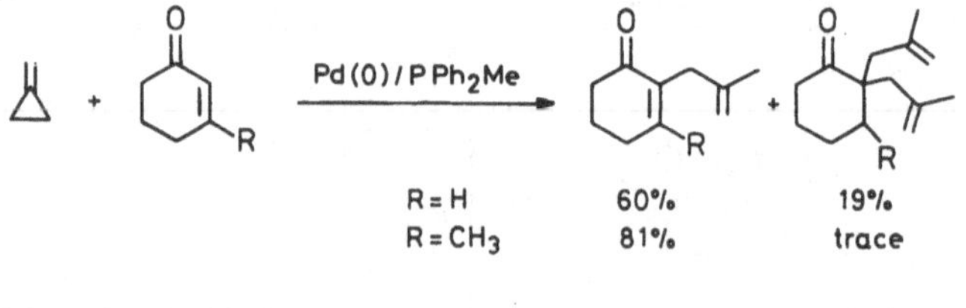

(84)

In our hands this reaction was more sluggish leading to a complex mixture of products including considerable amounts of phenol.

3. Only a limited number of vinylsulfones (e.g. *trans*-2-phenylvinyl-phenylsulfone) undergo codimerization with methylenecyclopropane. In most cases they are too strongly bound to the metal, preventing an interaction with methylenecyclopropane. In the case of alkyl-substituted vinylsulfones the Pd(0) catalyzed isomerization to allylsulfones is faster than cycloaddition [194].

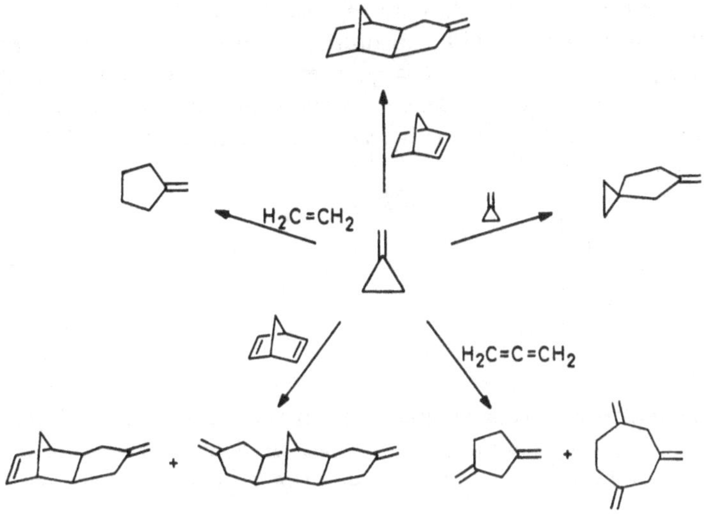

Scheme 7. Reactions of methylenecyclopropane with non-activated alkenes (catalyst: Pd(0)/P(*i*Pr)$_3$)

4. The two double bonds in 2,3-dimethoxycarbonylnorbornadiene are almost equally active. Furthermore, the reactions with methylenecyclopropane are stereoselective leading exclusively to the *exo*-isomers. Both observations are in striking contrast to the results obtained in the Pd(0) catalyzed cycloadditions of 2-[(trimethylsilyl)methyl]allyl acetate with norbornadiene derivatives [197].

The last observations automatically lead to the conclusion that non-activated alkenes also could undergo these reactions. Indeed it was found that ethylene, norbornene, norbornadiene [198] and allene [199] react with methylenecyclopropane to give cycloadducts (Scheme 7). The reason for the limitation to these alkenes lies in the ability of methylenecyclopropane to compete successfully with alkenes in π-complexation to the metal. Thus cyclodimerization of methylenecyclopropane is much faster than codimerization with other alkenes, which give less stable π-complexes with Pd(0).

This drawback can be overcome by the use of methylenecyclopropanes substituted at the vinylic position since tri- and tetrasubstituted alkenes show a much smaller tendency for π-complexation. In these cases, a number of 1-alkenes as well as cyclopentene (as the only example of a disubstituted alkene) and 1,3-butadiene react smoothly [200] (see pp. 125 and 127).

3.3.1.3 Codimerizations with Substituted Methylenecyclopropanes

For a broader application of these new metal catalyzed [3+2]-cycloadditions in organic syntheses it is necessary that substituted methylenecyclopropanes react chemoselectively, regioselectively and if possible stereoselectively in this manner. The codimerizations of methylenecyclopropane itself described above indicate that the course of these cycloadditions is extremly sensitive to a number of influences (see Sect. 3.3.1.1). It is therefore likely that the desirable selectivities of the codimerizations is much more difficult to achieve with substituted methylenecyclopropanes as e.g. with substituted 2-[(trimethylsilyl)methyl]allyl acetates. In the latter case it has been shown that, independent of the kind of substituent, all Pd(0) catalyzed cycloadditions lead to the same type of methylenecyclopentane derivative [183, 201] (Eq. 85).

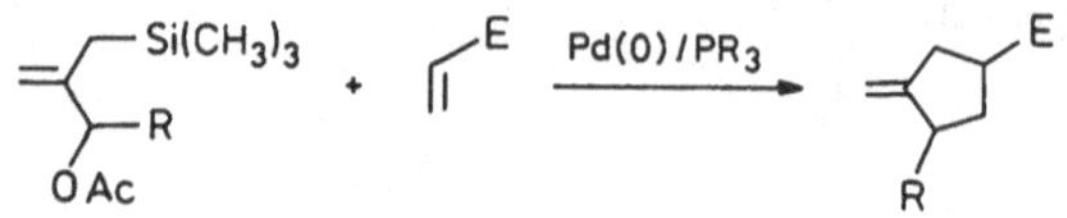

(85)

$R = CH_3$, Ph, $-CH{=}CH_2$, CN, $COCH_2CH_3$, OAc

With substituted methylenecyclopropanes the problem is even more complicated since in both types of [3+2]-cycloaddition presented in Scheme 5 two regioisomers can be formed. In Type a isomerization may occur in the "trimethylene-

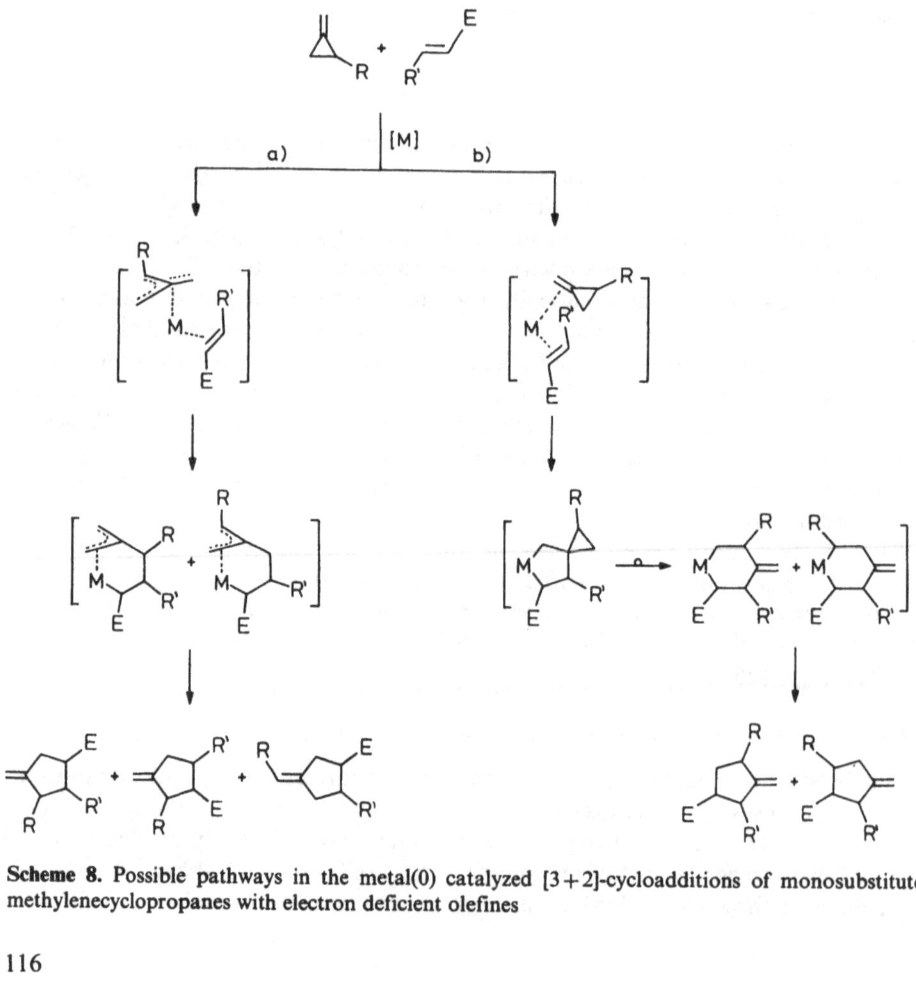

Scheme 8. Possible pathways in the metal(0) catalyzed [3+2]-cycloadditions of monosubstituted methylenecyclopropanes with electron deficient olefines

methane metal" intermediate, whereas in Type b the regioselectivity of the cycloaddition is determined by the selectivity of the cyclopropylmethyl/3-butenyl rearrangement (Scheme 8).

As will be shown in the following, all reaction modes outlined in Scheme 8 can be verified and especially in cycloadditions according to Path a it is difficult or impossible to obtain the desired regioselectivity of the reaction.

In principle one can distinguish between the reactivities of methylenecyclopropanes which are substituted at the three-membered ring and those which are substituted at the double bond. With Pd(0) catalysts, both types react in the same manner, by distal bond opening of the three-membered ring (Path a in Scheme 5 or 8) [202], whereas in the presence of Ni(0) catalysts the former types behaves like methylenecyclopropane itself, which means that they react according Path b in Scheme 5 or 8, if the substituent is an alkyl group. Substituents which are able to weaken the distal C—C bond of the three-membered ring induce also opening of this bond in the Ni(0) catalyzed reaction; e.g. with 2-phenylmethylenecyclopropane, codimerization following Pathway a and b of Scheme 8 can take place at the same time. With Ni(0) catalysts, general distal ring-opening is only observed when methylenecyclopropanes are used which are substituted at the double bond.

Therefore, in the presence of phosphane-free or phosphane-modified Ni(0) catalysts, alkyl acrylates and alkyl crotonates codimerize with 2-methyl- and 2,2-dimethylmethylenecyclopropanes to give Type B cycloadducts. The reactions turn out to be regio- but not stereoselective. The methyl group(s) bonded at the three-membered ring are always found at C-2 of the resulting methylenecyclopentanes, whereas the electron withdrawing group is bonded to C-4 (Eqs. 86 and 87) [27].

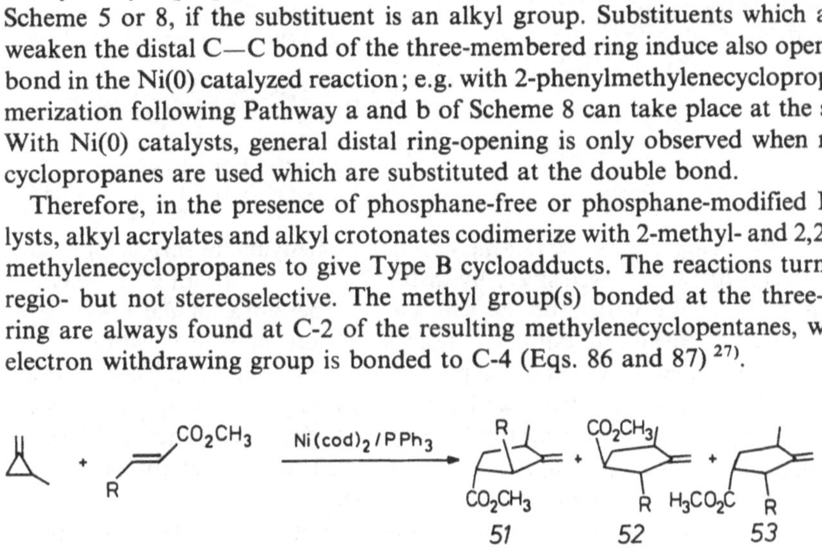

(86)

R	combined yield	ratio *51*:*52*:*53*
H	43%	70:30
CH_3	45%	55:25:20

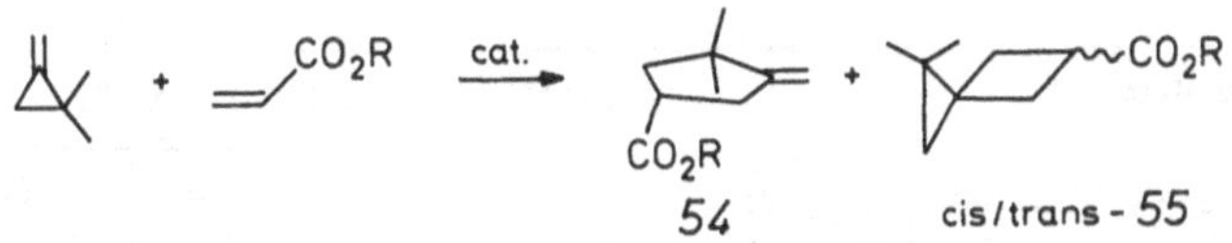

(87)

R	cat	combined yield	ratio *54*:*55*
CH_3	$Ni(cod)_2$ [175]	67%	81:19
CH_3	$Ni(an)_2$ [186]	60%	100: 0
CH_3	$Ni(cod)_2/PPh_3$ [27]	73%	92: 8
t-C_4H_9	$Ni(cod)_2$ [175]	68%	60:40

With 2,2-dimethylmethylenecyclopropane [2+2]-cycloaddition is also observed.

Interestingly, 2-phenylmethylenecyclopropane behaves differently. In the presence of phosphane-modified Ni(0) catalysts both types of cycloaddition products (A and B) are formed [27] (Eq. 88). Moreover the regioselectivity of Pathway b changes, as considerable amounts of *trans*-3-phenyl-4-methoxycarbonyl-methylenecyclopentanes are formed. On the other hand, the stereoselectivity of cycloaddition Products B is much better than in the same reactions with 2-methylmethylenecyclopropane.

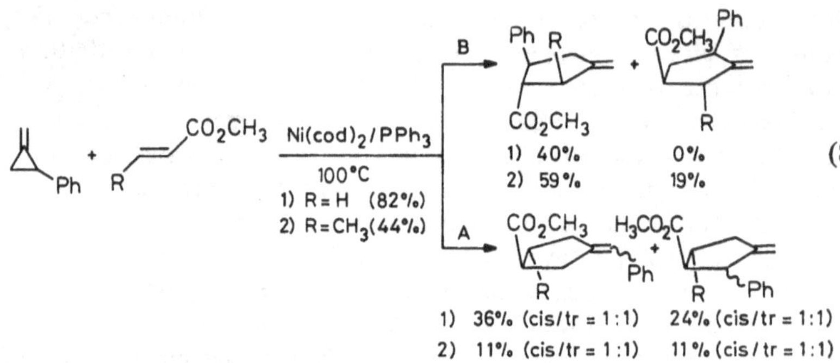

(88)

Like methylenecyclopropane itself, 2-methyl- and 2-phenylmethylenecyclopropane react with dialkyl fumarates exclusively by distal ring-opening (Path a in Scheme 5 or 8) independent of the nature of the catalyst [27]. Therefore, in the presence of phosphane-modified Ni(0) and Pd(0) catalysts a mixture of ring substituted and double bond substituted methylenecyclopentanes is obtained, the phenyl group favoring the latter. However, with 2-phenylmethylenecyclopropane the product distribution can be changed favoring the 2-phenylmethylenecyclopentane derivative by 88:12 using a Pd(0)/P(*i*Pr)$_3$ catalyst with a ratio 1:4. The same cannot be achieved with the less reactive 2-methylmethylenecyclopropane.

On the other hand, the latter compound is isomerized to some extent to isoprene using a Pd(0)/P(*i*Pr)$_3$ catalyst, and that undergoes Diels-Alder cycloadditions under the reaction conditions.

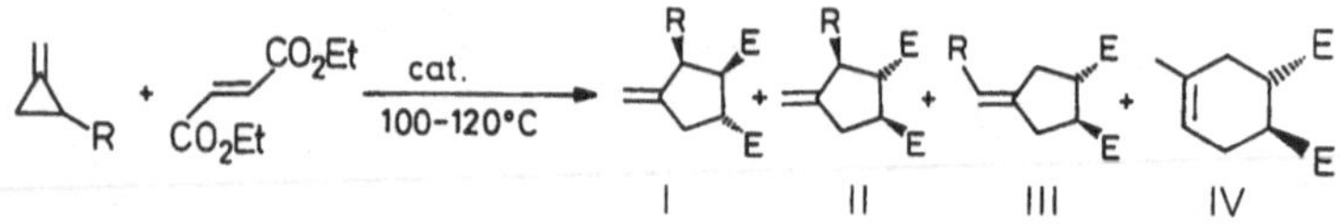

(89)

R	catalyst	yield [%]	composition [%]			
			I	II	III	IV
CH_3	Ni(cod)$_2$/1 PPh$_3$	70	49	6	45	—
CH_3	Ni(cod)$_2$/1 TOPP	63	40	21	35	4
CH_3	Ni(cod)$_2$/4 TOPP	57	37	28	30	5
Ph	Ni(cod)$_2$/1 PPh$_3$	93	4	10	86	—
CH_3	cpPd(allyl)/1 P(*i*Pr)$_3$	83	20	12	48	20
Ph	cpPd(allyl)/1 P(*i*Pr)$_3$	85	13	16	71	—
Ph	cpPd(allyl)/4 P(*i*Pr)$_3$	86	49	39	12	—

The advantage of using methylenecyclopropanes with substituents at the double bond is, that they react with both Ni(0) and Pd(0) based catalysts in the same way: by distal opening of the three-membered ring. Moreover, the substituents at the double bond seem to relieve the distal ring-opening and to hinder a cyclodimerization of the methylenecyclopropane itself. The results of these changes in properties are that the metal catalyzed codimerizations can be carried out under milder conditions and that some less reactive alkenes, which do not undergo [3+2]-cycloadditions with methylenecyclopropane or ring substituted methylenecyclopropanes, do react in this manner with double bond substituted species [203, 204]. Examples for this behavior are the successful annelation of 2-cyclohexenone and 2-cycloheptenone with isopropylidenecyclopropane (Table 6) and (diphenylmethylene)cyclopropane (Table 9) and the codimerization of 1-alkenes, cyclopentene and some 1,3-butadienes with the same methylenecyclopropanes (Tables 7 and 10).

Nevertheless, some characteristic differences between the Ni(0) and the Pd(0) catalysts exist. The most important ones are summarized in the following. For more details the reader should compare the experimental results listed in Tables 4–10.

The tendency to give 2-substituted methylenecyclopentanes is much higher in the presence of phosphorous-modified Ni(0) catalysts. With triisopropylphosphane/Pd(0) catalysts nearly always alkylidenecyclopentanes are formed and often exclusively. The formation of 2-substituted methylenecyclopentanes can be achieved using phosphite containing Ni(0) catalysts with bulky aryl groups. A higher P:Ni-ratio also favours the formation of these isomers. Besides the nature of the catalyst, the properties of the olefines also influence the ratio of 2-substituted methylenecyclopentanes to alkylidenecyclopentanes. The latter are exclusively formed with alkenes. With olefines bearing electron withdrawing groups, very subtile influences, which are not yet fully understood, play a role; e.g. with methyl acrylates or acrylonitrile alkylidenecyclopentanes are formed mainly, independent of the type of the catalyst. On the other hand with crotonates or fumarates 2-substituted methylenecyclopentanes can be synthesized with up to 90% yield (see Tables 4 and 6).

Another interesting point is the regioselectivity of these [3+2]-cycloadditions. Whenever a 2-substituted methylenecyclopentane is detected in a codimerization, catalyzed by a Pd(0) compound, the substitution pattern is the same as found in [3+2]-cycloadditions starting with 2-[(trimethylsilyl)methyl]allyl acetates [183] (Eq. 85).

In contrast to this all phosphorous-modified Ni(0) catalysts lead to another regioselectivity in which the electron withdrawing group is bonded in the neighborhood to the substituent of the methylenecyclopropane (Eq. 90).

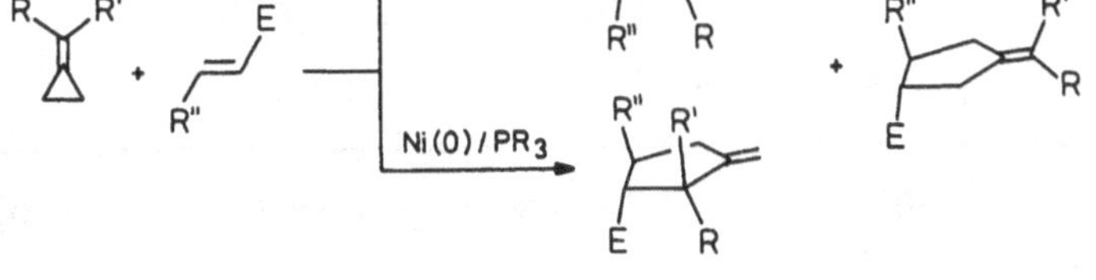

(90)

Though this different behavior of Pd(0) and Ni(0) catalysts is not yet fully understood, it may serve to complement the possibilities of synthesizing regioselectively substituted methylenecyclopentanes.

Table 4. Codimerization of hexylidenecyclopropane with electron deficient olefines

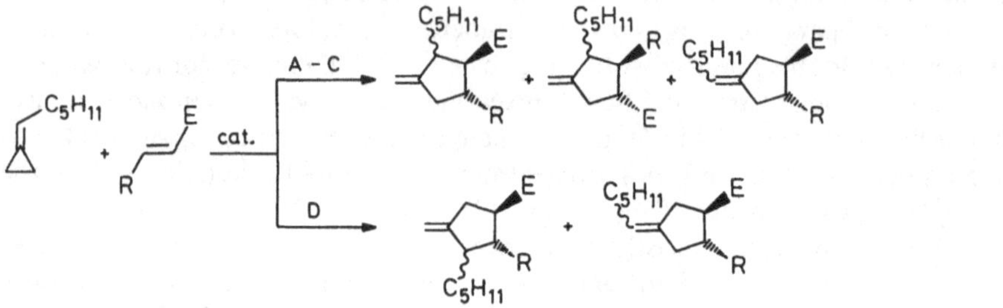

(91)

Olefine	Cat.	Temp. [°C]	Yield [%]	Codimers (*cis* : *trans* ratio)		
				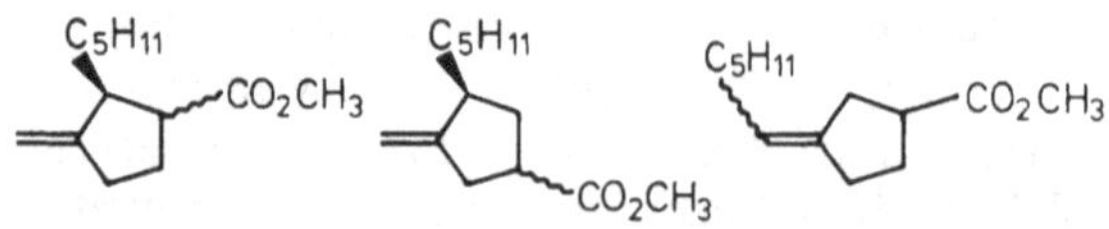		
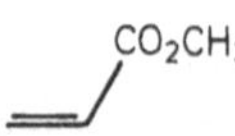	A	100	78	10 (0:100)	24 (1:1)	66 (1: 2)
	B	120	72	23 (78: 22)	43 (1:1)	34 (23:77)
	C	120	80	31 (9: 1)	38 (1:1)	31 (22:78)
	D	130	53	0	8 (1:1)	92 (4: 6)

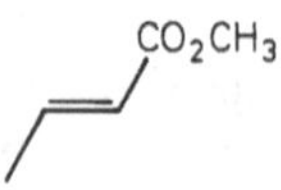	C	120	34	100 (12: 88)		0
	D	130	51	0		100
				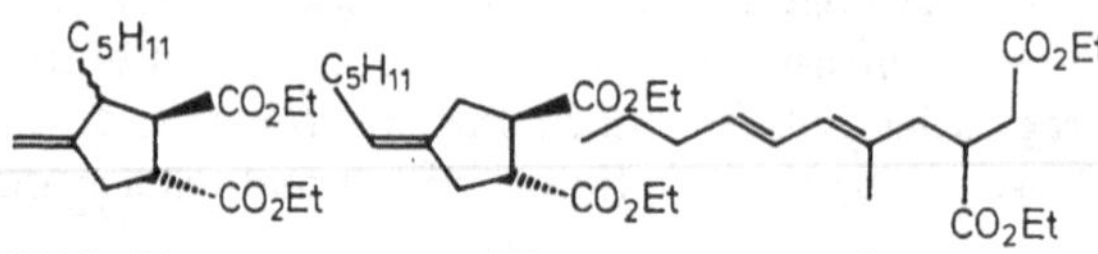		
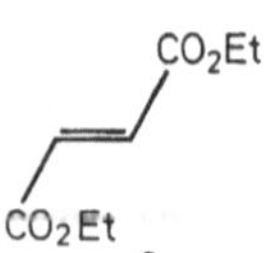	A	120	87	43 (8:2)	57	0
	B	120	21	62 (6:4)	38	0
	C	120	75	81 (4:6)	19	0
	D	130	72	37	43	20
					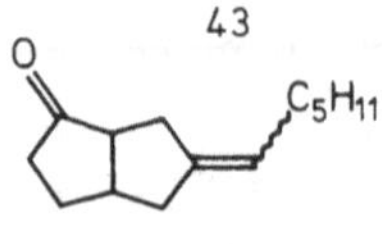	
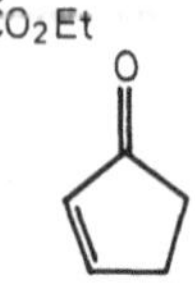	A	120	0			
	D	120	61		100 (1:1)	

Catalysts: A = Ni(cod)$_2$ + 1 PPh$_3$; B = Ni(cod)$_2$ + 1 TOPP; C = Ni(cod)$_2$ + 4 TOPP; D = cpPd(allyl) + 1 P(*i*Pr)$_3$

Table 5. Codimerizations of (phenylmethylene)cyclopropane with electron deficient olefines

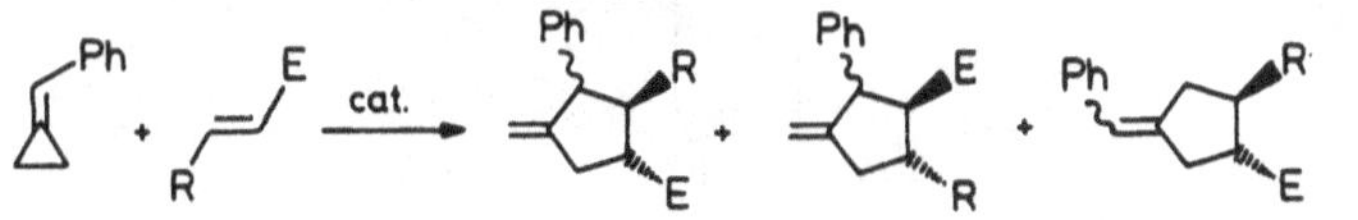

(92)

Olefine	Cat.	Temp. [°C]	Yield [%]	Codimers (*cis* : *trans* ratio)		
				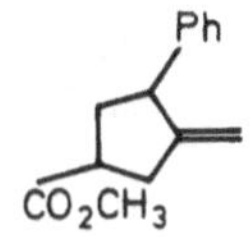	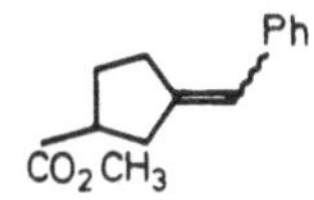	
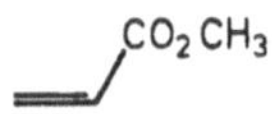	A	50	92	0	100 (3 : 2)	
	B	100	94	0	100 (45 : 55)	
	C	100	88	8	92 (58 : 42)	
				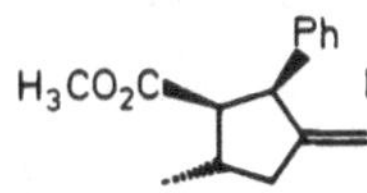	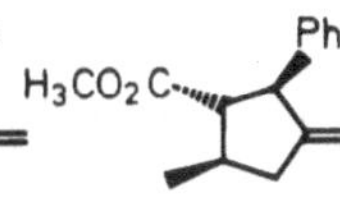	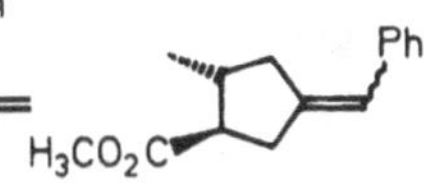
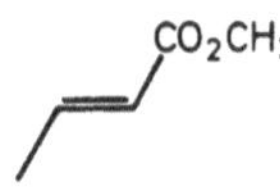	B	90	23	55	0	45
	C	100	52	0	0	100 (2 : 1)
				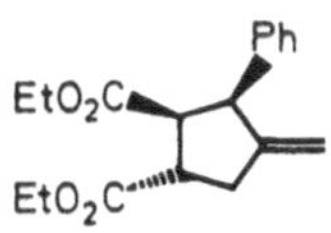	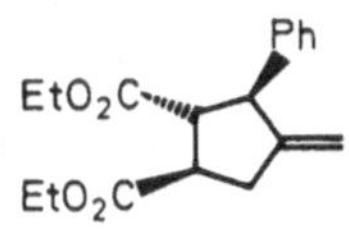	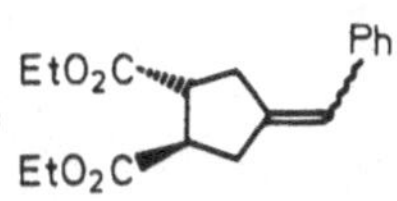
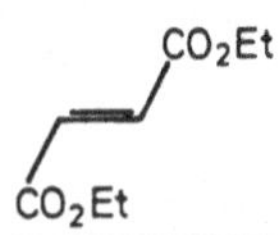	B	90	82	4	4	92
	D	140	67	30	25	45

Catalysts: A = $Ni(cod)_2$; B = $Ni(cod)_2$ + 1 PPh_3; C = cpPd(allyl) + 1 $P(iPr)_3$; D = cpPd(allyl) + 4 $P(iPr)_3$

3.3.2 Methylenecyclopentenes

Whereas the transition metal catalyzed cyclotrimerization and cyclotetramerization of alkynes leading to benzene or cyclooctatetraene and their derivatives is a rather common reaction, there exist only a few examples of cooligomerizations between alkynes and alkenes or 1,3-butadienes leading to 1,3- or 1,4-cyclohexadiene derivatives [205]. It is therefore surprising that the [3+2]-cycloaddition between methylenecyclopropanes and alkynes, catalyzed by triarylphosphite modified Ni(0) compound, is a rather convenient method to synthesize 4-methylenecyclopentenes [206]. A wide range of methylenecyclopropanes and alkynes, in the latter case mainly 1,2-disubstituted ones, can be used for these reactions (Eqs. 98–100, see p. 127–128).

Table 6. Codimerizations of isopropylidenecyclopropane with electron deficient olefines

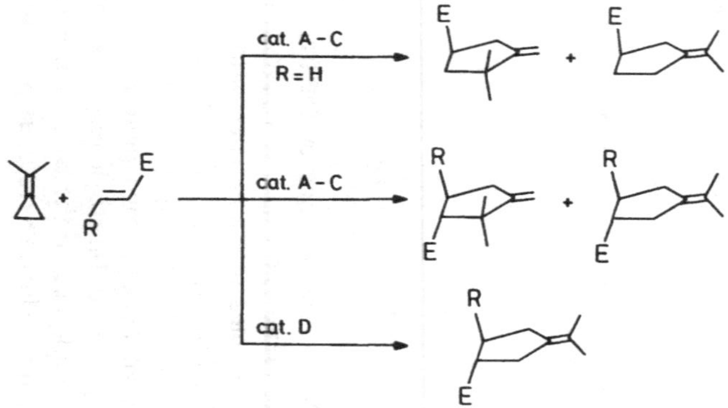

(93)

Olefine	Cat.	Temp. [°C]	Yield [%]	Codimers	
				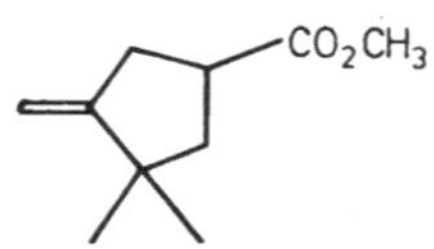	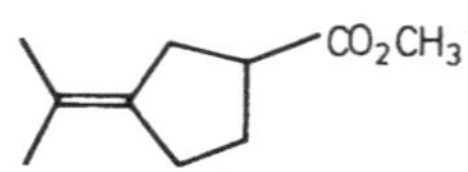
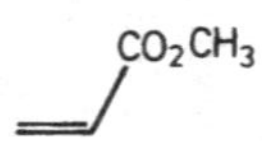	A	100	59	4	96
	B	100	89	15	85
	C	100	74	15	85
	D	130	48	0	100

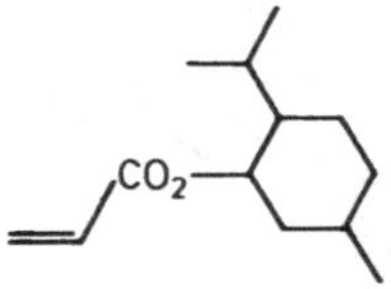	B	120	88	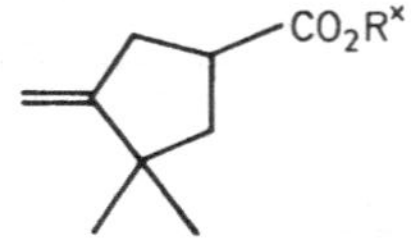 37 (d.e.0)	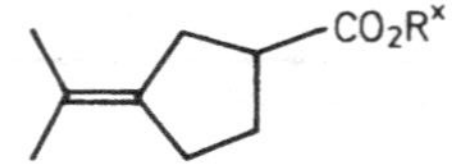 63 (d.e.10)	
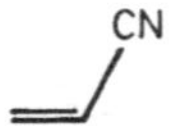	B D	120 130	51 0	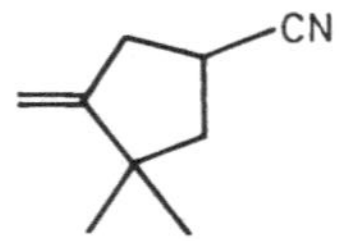 18 —	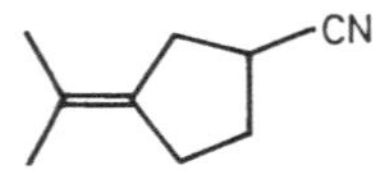 82 —	
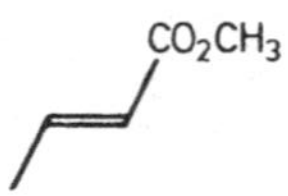	A B C D	100 120 120 130	59 44 46 71	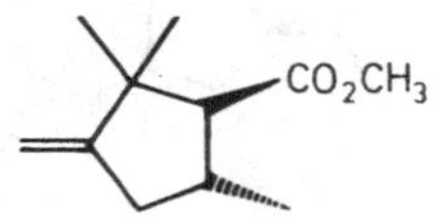 29 82 89 0	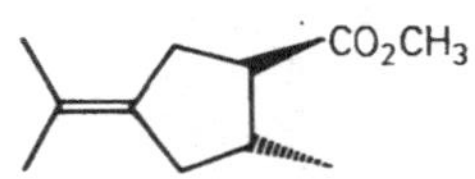 52 18 11 100	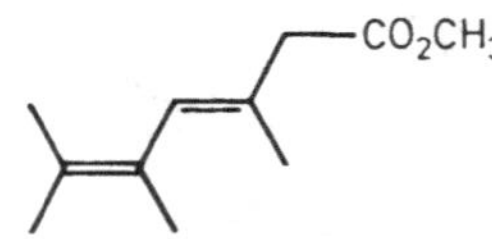 19 0 0 0
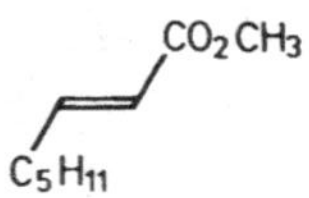	D	130	61	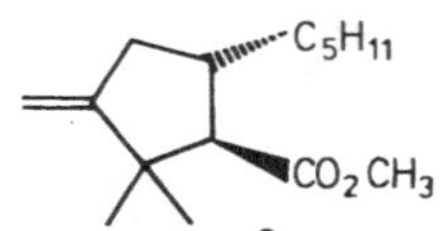 0	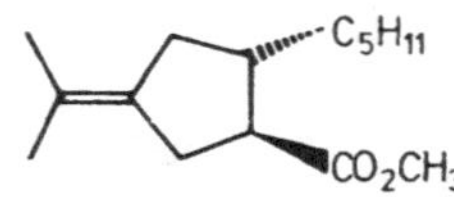 100	
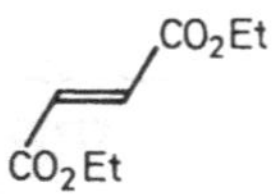	B C D	120 120 130	85 88 88	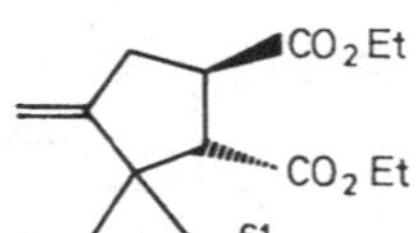 61 60 0	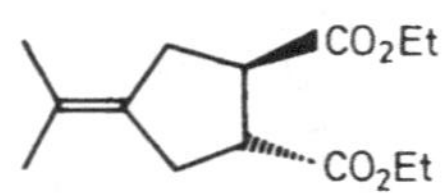 37 38 94	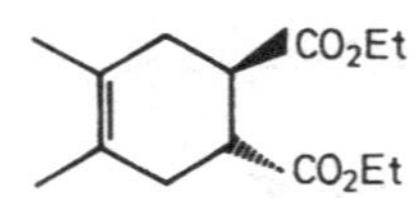 2 2 6

Olefine	Cat.	Temp. [°C]	Yield [%]	Codimers			
				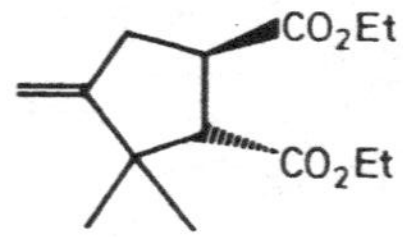	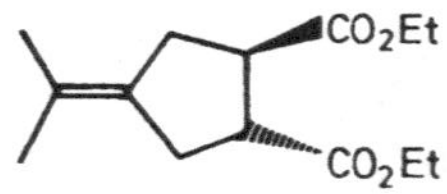	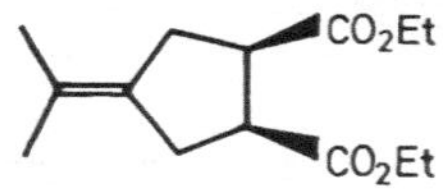	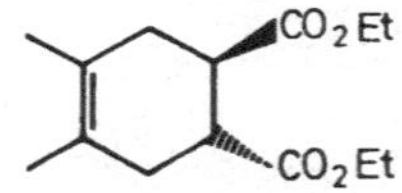
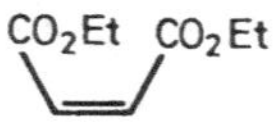	B D	130 130	90 53	28 1	63 57	0 29	9 13
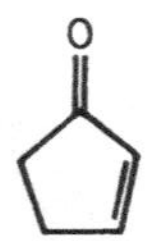	B D	120 100	0 71			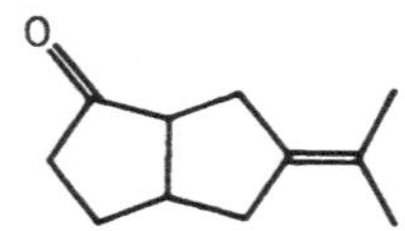 — 100	
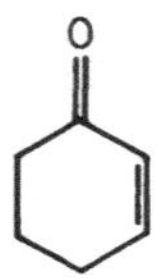	A D	120 130	10 68	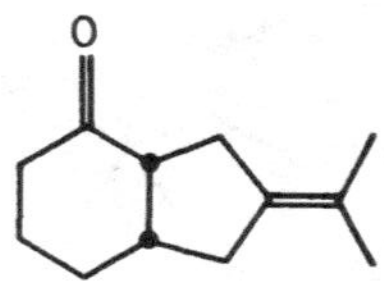 0 80		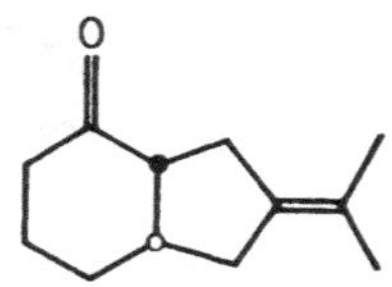 0 20	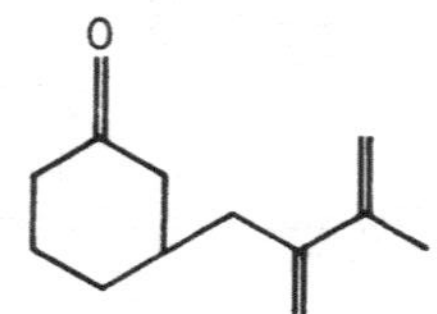 100 0
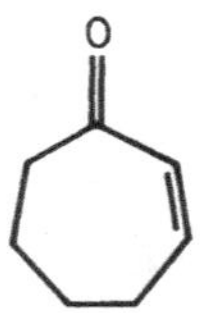	B D	120 130	57 50	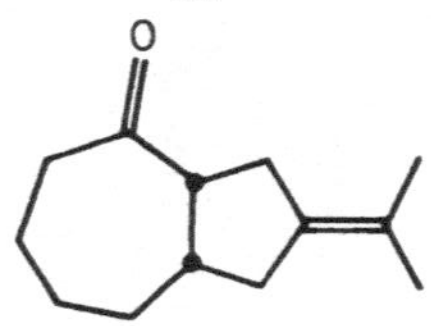 0 11		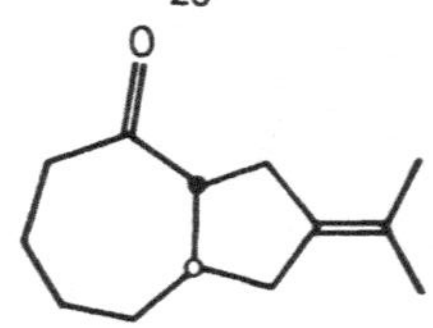 0 81	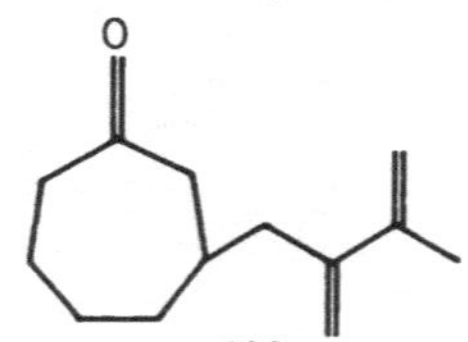 100 8

Catalysts: A = $Ni(cod)_2$ + 1 PPh_3; B = $Ni(cod)_2$ + 1 TOPP; C = $Ni(cod)_2$ + 4 TOPP; D = cpPd(allyl) + 1 $P(iPr)_3$

Table 7. Codimerizations of isopropylidenecyclopropane with alkenes

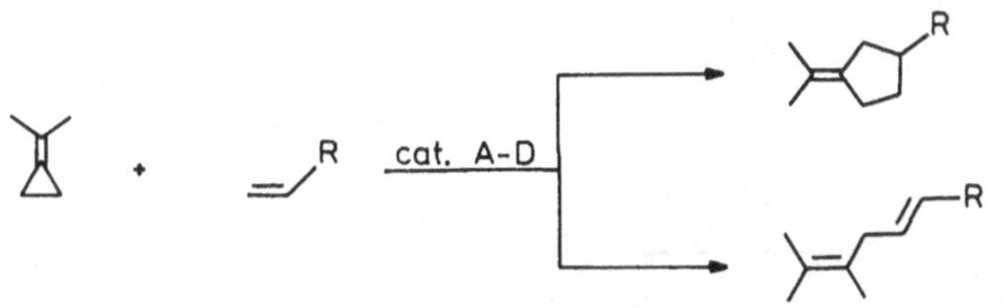

(94)

Alkene	Cat.	Temp. [°C]	Yield [%]	Codimers		
				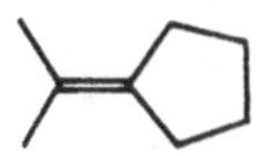	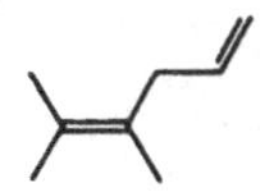	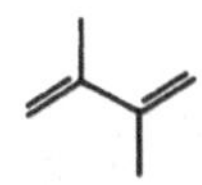
$CH_2{=}CH_2$	A	60	80	100	0	0
	C	120	82	100	0	0
	D	110	96	1	13	76
				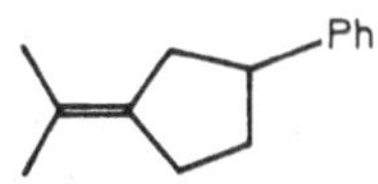		
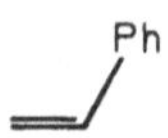	A	90	77	100		
	C	90	87	100		
				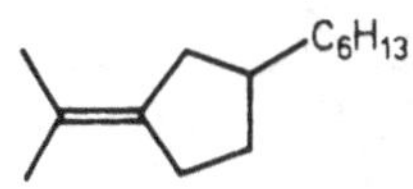	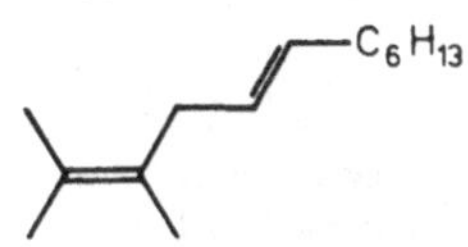	
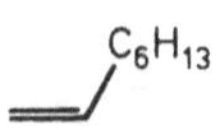	A	80	0	–	–	
	B	100	48	82	18	

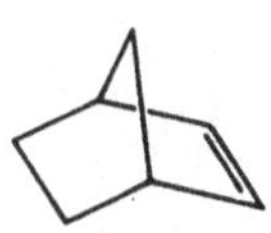	A	60	90	100		
	D	100	80	100		
				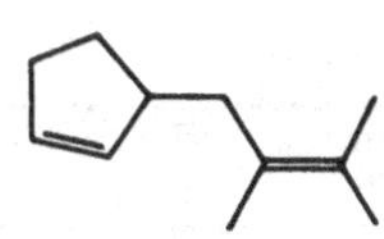		
	A	80	0	–		
	B	100	42	100		

Catalysts: A = Ni(cod)$_2$; B = Ni(cod)$_2$ + 1 PPh$_3$; C = Ni(cod)$_2$ + 1 TOPP; D = cpPd(allyl) + 1 P(*i*Pr)$_3$

Table 8. Codimerizations of (diphenylmethylene)cyclopropane with electron deficient olefines

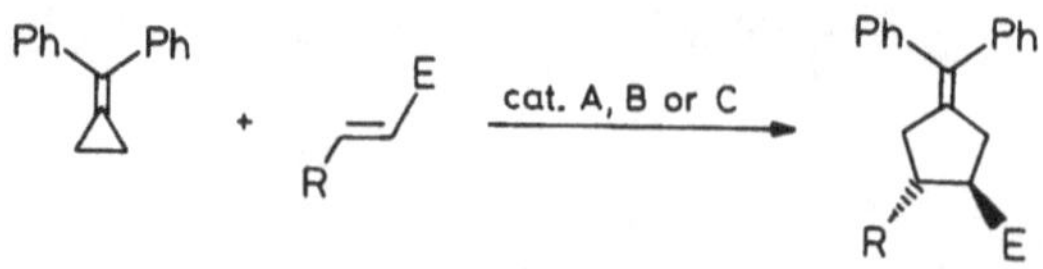

(95)

R	E	Cat.	Temp. [°C]	Yield [%]	Remarks
H	CO_2CH_3	A	70	79	
H	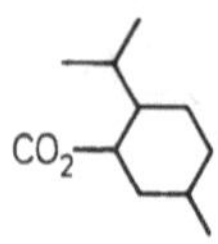	A	25	45	d.e. = 0
H	CO_2CH_3	C	110	86	
CH_3	CO_2CH_3	B	120	75	
CH_3	CO_2CH_3	C	110	94	
C_6H_{13}	CO_2CH_3	C	120	98	
H	CN	B	120	82	
CO_2Et	CO_2Et	B	100	98	
CO_2Et	CO_2Et	C	100	100	

Catalysts: A = $Ni(cod)_2$; B = $Ni(cod)_2$ + 1 TOPP; C = cpPd(allyl) + 1 $P(iPr)_3$

Table 9. Codimerizations of (diphenylmethylene)cyclopropane with 2-cycloalkenones

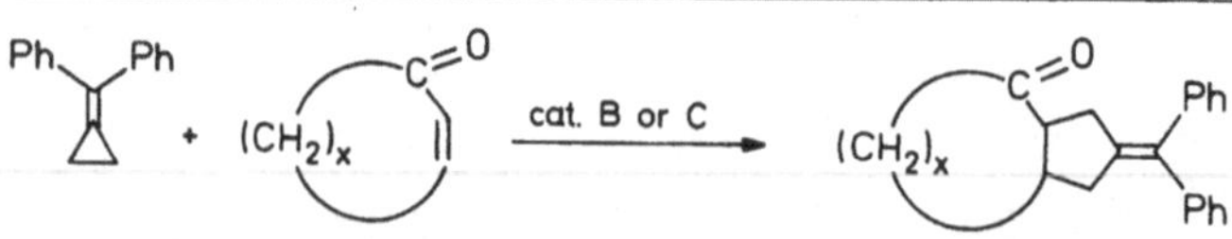

(96)

x	Cat.	Temp. [°C]	Yield [%]	*cis*	*trans*
2	B	120	0	–	–
2	C	120	97	100	–
3	B	120	0	–	–
3	C	125	76	76	24
4	C	125	80	16	84

Catalysts: B = $Ni(cod)_2$ + 1 TOPP; C = cpPd(allyl) + 1 $P(iPr)_3$

Table 10. Codimerizations of (diphenylmethylene)cyclopropane with alkenes

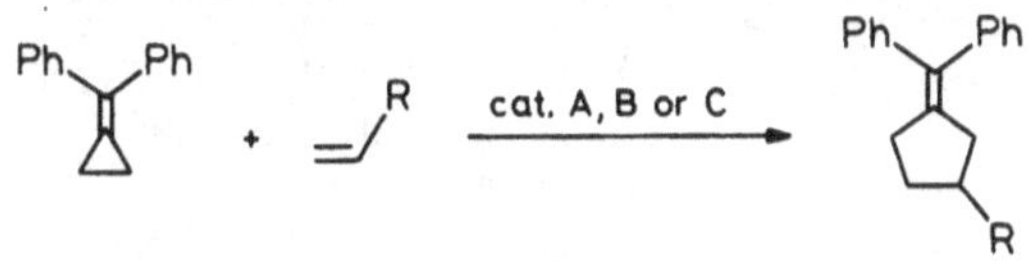

(97)

R[alkene]	Cat.	Temp. [°C]	Yield [%]	Other products	
H	A	80	37		
H	C	100	83		
CH_3	A	90	15		
CH_3	C	100	83		
C_4H_9	B	120	80		
C_4H_9	C	100	85		
$(CH_2)_8CO_2CH_3$	B	120	28		
$(CH_2)_8CO_2CH_3$	C	100	50		
Ph	C	100	88		
$CH{=}CH_2$	C	120	48		
$CH{=}CPh_2$	A	25	79		
	B	120	70	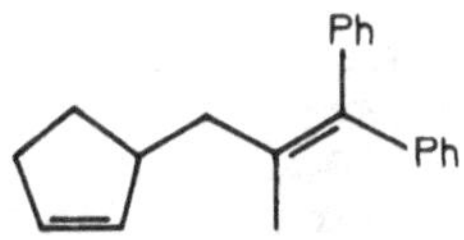	(100%)
	C	130	61	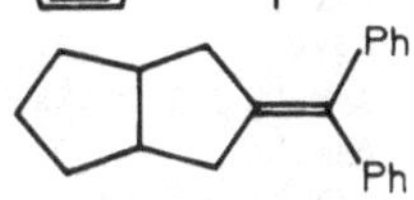	(100%)
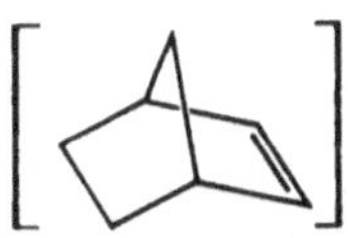	A	50	70	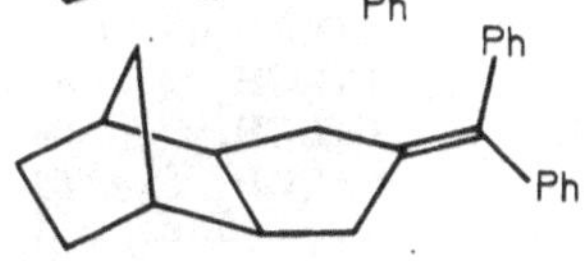	(100%)
	C	70	94		(100%)

Catalysts: A = $Ni(cod)_2$; B = $Ni(cod)_2$ + 1 TOPP; C = cpPd(allyl) + 1 $P(iPr)_3$

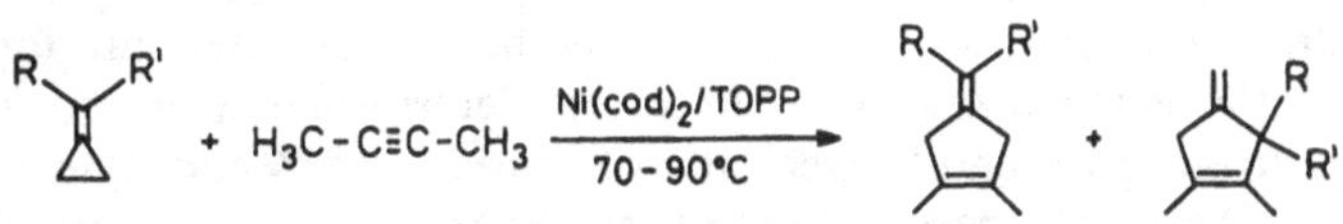

64 *65*

(98)

R	R'	yield [%]	*64* : *65* ratio	
H	H	34	100	
CH_3	CH_3	66	6	94
H	Ph	57	0	100
Ph	Ph	95	6	94

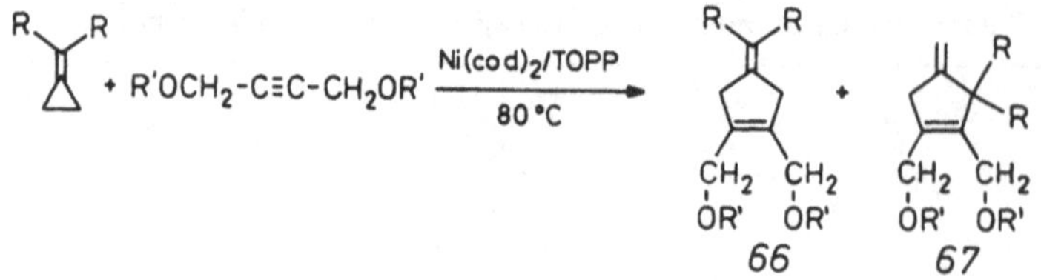

(99)

R	R′	Yield [%]	66	67
H	$SiMe_3$	35	100	
H	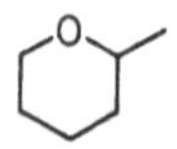	52	100	
Ph	$SiMe_3$	86	57	43

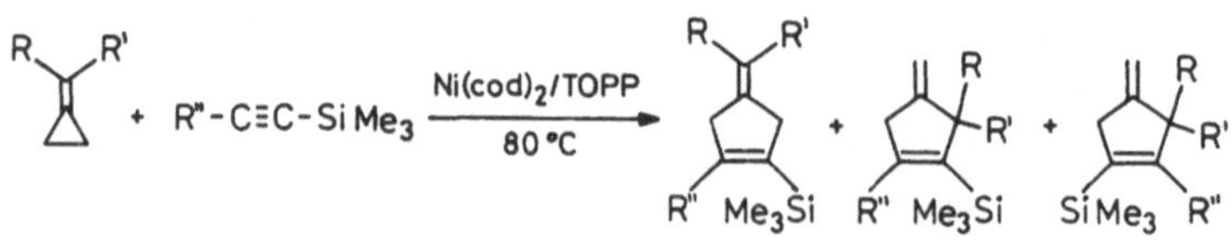

(100)

R	R′	R″	yield [%]	*68*	*69*	*70*
CH_3	CH_3	$SiMe_3$	50	0		100
Ph	Ph	$SiMe_3$	71	48		52
H	H	Alkyl	48			
H	nC_5H_{11}	Alkyl	82	12	63	25
CH_3	CH_3	Alkyl	88	0	85	15
Ph	Ph	Alkyl	92	60	24	16
H	H	CO_2CH_3	72			
H	nC_5H_{11}	CO_2CH_3	65	25	22	53
CH_3	CH_3	CO_2CH_3	64	41	21	38
Ph	Ph	CO_2CH_3	63	100	0	0

It is characteristic for these cyclodimerizations that all types of methylenecyclopropanes react in the same way: by distal ring-opening. Unfortunately in many cases the regioselectivity of the cycloaddition is not very high. Especially when unsymmetrically disubstituted alkynes are used, both possible regioisomers are formed with a preference of the isomers in which the more electronegative group is positioned neighboring the groups R and R′ (see ratio of compounds *69* and *70* in Eq. 100).

The resulting 4-methylenecyclopentenes are very sensitive to base- or acid-catalyzed isomerization of the exocyclic double bond into the ring, yielding 1,3-cyclopentadienes; e.g. it was not possible to substitute the TMS-group by hydrogen with the aid of trifluoroacetic acid without isomerization.

3.3.3 Heterocycles

As highly reactive olefines methylenecyclopropanes should be predestinated to undergo thermal or transition metal catalyzed cyclocodimerizations with CX triple bonds or double bonds (X = O, N). In practice only few examples of such cycloadditions with heterocumulenes, like ketenes, keteneimines, carbodiimides or carbondioxide are known.

This behaviour correponds to the observation that other unsaturated hydrocarbons, e.g. alkynes, allenes or 1,3-butadienes, which readily undergo transition metal catalyzed cyclooligomerizations, do also incorporate CX multiple bonds in such cycloadditions only with difficulty in most cases [207, 208]. Besides the well known cobalt-catalyzed pyridin synthesis from alkynes and nitriles [98] cocyclooligomerizations have been achieved with alkynes on one side and isocyanates [209], carbodiimides [210] and carbondioxide [211] on the other side as well as with 1,3-butadienes and aldehydes [212], carbondioxide [213] and 2-aza- or 2,3-diaza-1,3-butadiene [214].

In the thermally induced [2+2]-cycloaddition of methylenecyclopropanes with ketenes only the C=C-double bond of the ketenes react to give spiro[3.2]hexanones in good yields [215]. The reaction proceeds stereospecificly with methylenecyclopropanes substituted at the double bond. With diphenylketene, tetrahydronaphthalene derivatives are also formed [216]. Transition metal compounds do not catalyze these or other reactions between methylenecyclopropanes and ketenes.

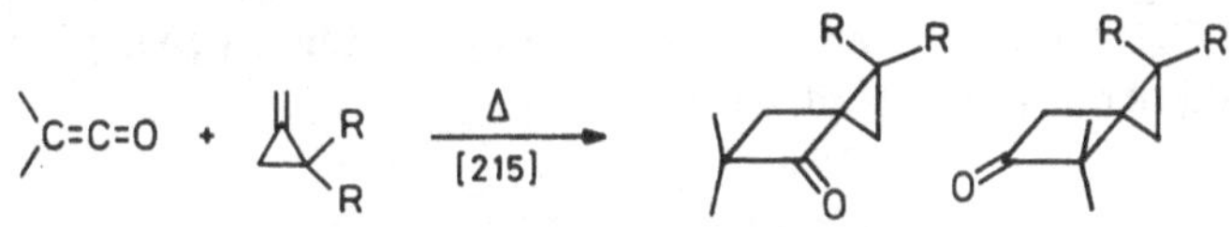

(101)

R = H; CH_3

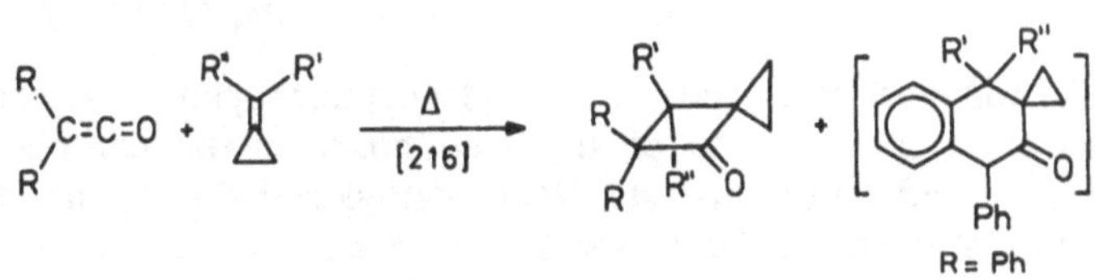

(102)

R = Et, Ph, Cl
R' = H, CH_3, Ph, $-(CH_2)_5-$
R'' = CH_3, Ph,

In contrast to ketenes, keteneimines readily undergo metal-catalyzed [3+2]-cycloadditions with methylenecyclopropanes. Depending on the substituents in both educts, pyrroles, α-methylene-Δ^3-pyrrolines or iminocyclopentenes can be synthesized selectively in good to excellent yields. The substituent of the imino group determines

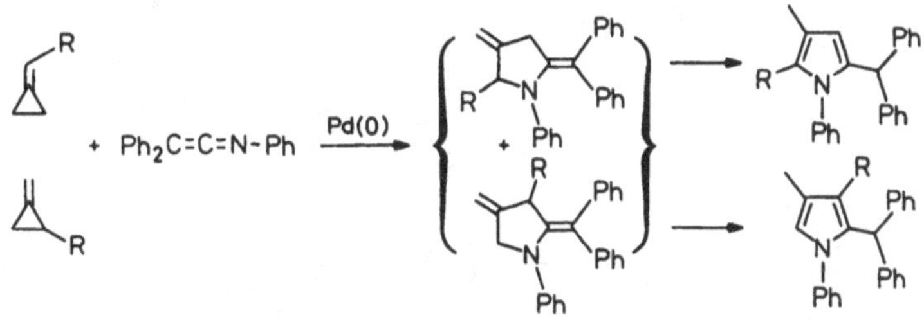

(103)

R = H, CH_3

which double bond of the keteneimine reacts whereas the substituents on the methylenecyclopropanes influences the stabilization of the primary cycloadduct. With triphenylketeneimine, all methylenecyclopropanes react by addition to the imino double bond.

The primary cycloadducts, 2,4-dimethylene-N-phenyl-pyrrolidines, normally isomerize to pyrroles during the reaction (Eq. 103). Only with geminally disubstituted methylenecyclopropanes, α-methylene-Δ^3-pyrrolines are formed (Eq. 104) [217].

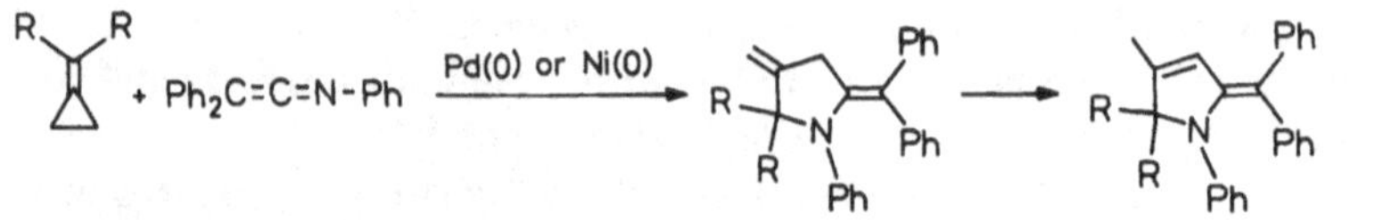

(104)

R = CH_3, (CH.).

For all these codimerizations Pd(0) as well as Ni(0) catalysts can be used, preferred catalysts are $Pd(PPh_3)_4$ or mixtures of $Ni(cod)_2$ and a arylphosphite in a molar ratio of 1:1. The less reactive diphenylketene-N-methylimine only reacts with methylenecyclopropane to give cycloadducts in moderate yields [catalyst: $Pd(PPh_3)_4$]. Interestingly only the C=C double bond undergoes the cycloaddition giving 1-methylimino-3-methyl-5,5-diphenyl-2-cyclopentene [217].

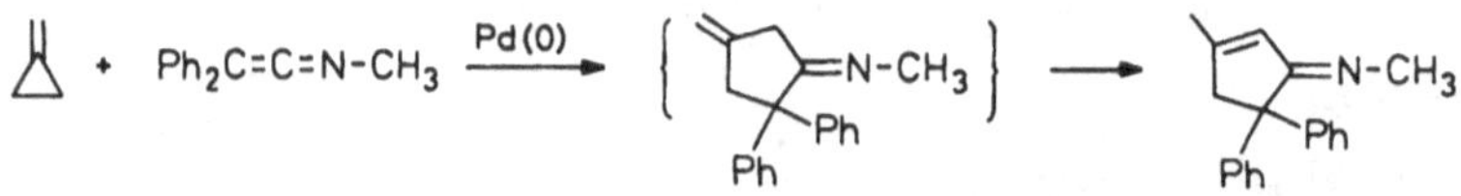

(105)

A successful codimerization between methylenecyclopropanes and carbondioxide is only possible in the presence of triphenylphosphane or tetraphenylethylenediphosphane modified Pd(0) catalysts. With methylenecyclopropane itself, a high yield preparation of the butenolid *71* is somewhat tricky, because *71* contains at least three acidic hydrogens, which can react further with methylenecyclopropane. Therefore *71* is obtained in 80% yield only under special reaction conditions, e.g. by pumping a solution of the catalyst and the methylenecyclopropane in DMF slowly into a 165 °C hot autoclav which contains the same solvent and is under a pressure of 40 bar CO_2

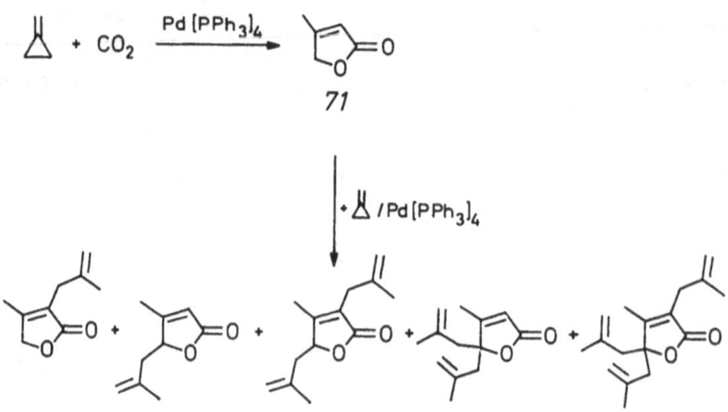

71

(106)

during the whole reaction [218]. Otherwise cooligomers, as shown in Eq. 106, are also formed in more or less quantities.

Codimerization between substituted methylenecyclopropanes and CO_2 are reported to be less complicated [219]; e.g. isopropylidenecyclopropane or butylidenecyclopropane are published to give the corresponding butenolids in 77% or 67% yield by simple heating a benzene solution of these compounds to 130 °C in the presence of $Pd(PPh_3)_4$ as a catalyst and under a pressure of 40 bar CO_2 (Eqs. 107 and 108). But it should be noted that in our hand a reaction with an excess of the methylenecyclopropane could not be avoided leading to cooligomeres analogous to those formulated in Eq. 106 [218].

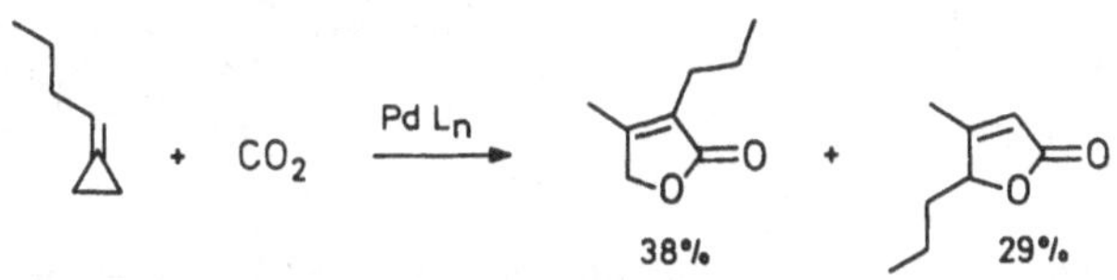

(107)

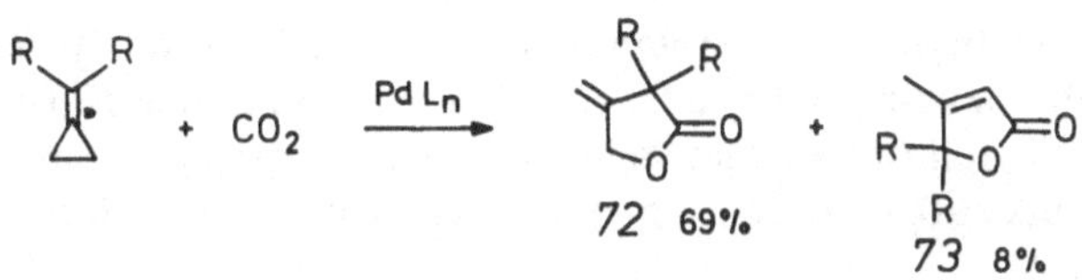

Under prolonged heating to 125 °C the furanone *72* can be isomerized to the butenolid *73* under the influence of a Pd(0) catalyst [220].

3.4 Synthesis of Seven-Membered Rings

Two strategies are available for the synthesis of seven-membered carbocycles starting with methylenecyclopropanes:

1) [3+2+2]-cycloadditions (Eq. 109)
2) [4+3]-cycloadditions (Eq. 110)

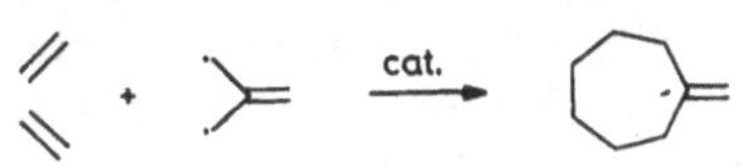

(109a)

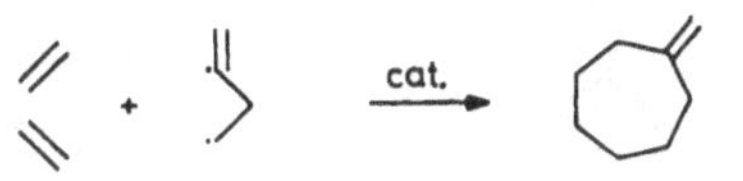

(109b)

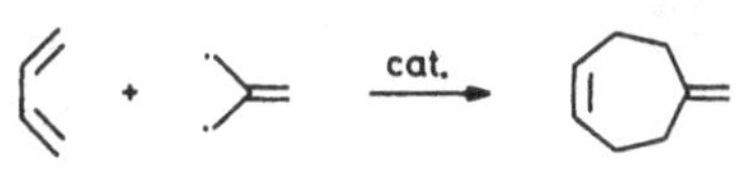

(110)

These cycloadditions can be achieved only with the help of a Ni(0) or Pd(0) catalyst. Moreover, only one example of each reaction has been observed until now.

A [3+2+2]-cycloaddition according to Eq. 109a has been achieved only in the case of methylenecyclopropane and allene, which react smoothly in the molar ratio of 1:2 in the presence of a $(iPr)_3P/Pd(0)$ catalyst to give 1,3,6-trimethylenecycloheptane in good yields. 1,3-Dimethylenecyclopentane and a cyclodimer of methylenecyclopropane are formed as by-products [199].

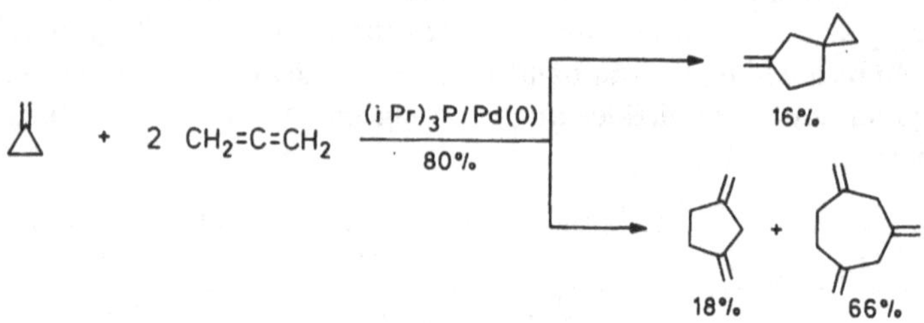

(111)

Substituted allenes, e.g. 1,1-dimethylallene do not react in the same manner, but rather lead to codimers [199].

Other types of methylenecycloheptanes are obtained by the trimerization of methylenecyclopropane catalyzed by a trialkylphosphane modified Ni(0) complex [221]. Although most trialkylphosphan ligands induce the formation of open-chain trienes, phosphanes with bulky alkyl groups, e.g. *tert* butyl, produce methylenecycloheptanes in reasonable yields.

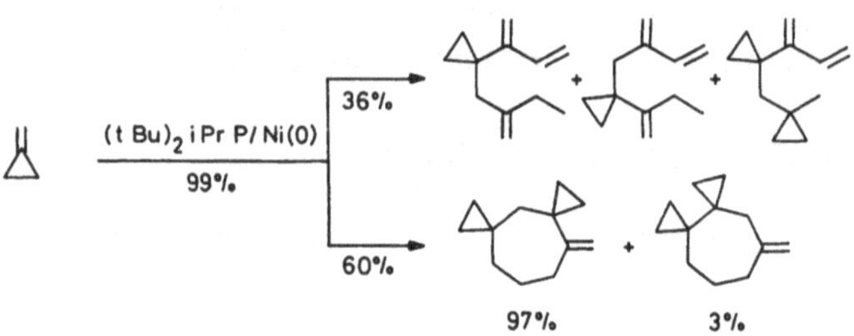

(112)

It should be pointed out that these [3+2+2]-cycloadditions are not metal induced electrolytic processes but stepwise reactions in which nickelacyclopentanes (Eq. 112) or nickelacyclohexanes (Eq. 111) are initially formed, followed by insertion of one of the M—C-bond into the C=C double bond of the second olefin.

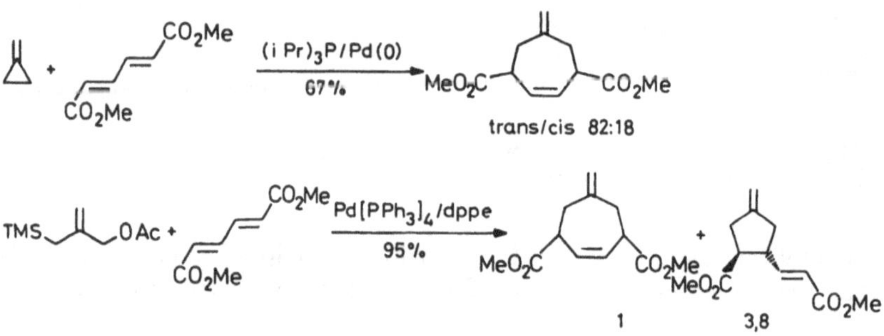

(113)

[4+3]-cycloaddition according Eq. 110 can be carried out with dimethyl (*E*, *E*) muconate and methylenecyclopropane, which codimerize in the presence of a Pd(0) catalyst to give a seven-membered ring product as a 82:18 mixture of the trans-cis-isomers in 67% yield [222] (Eq. 113). In contrast to the results obtained with 2-[(trimethylsilyl)methyl]allyl acetate [223], no vinylcyclopentane derivative could be detected.

Other 1,3-dienes, such as 1,3-butadiene, isoprene and methyl-2,4-pentadienoate, either do not react with methylenecyclopropanes or yield only 3-vinylmethylenecyclopentane derivatives exclusively (Table 10 and Eq. 114). Quite unexpectedly, methyl-2,4-pentadienoate reacts only at the terminal C=C bond, giving a vinylmethylenecyclopentane in poor yield [224] (Eq. 114).

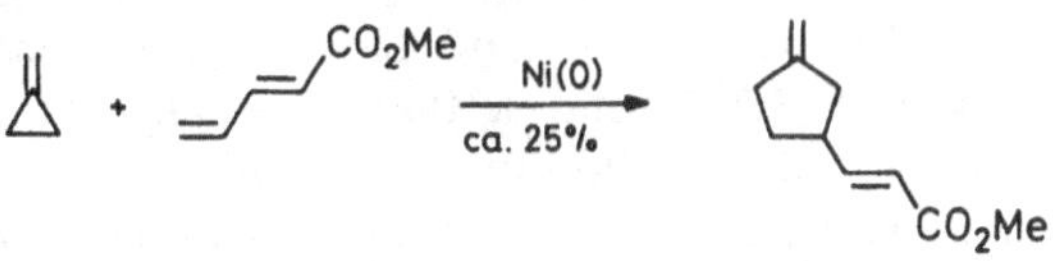

(114)

3.5 Some Applications of the Metal Catalyzed [3+2]-Cycloaddition with Methylenecyclopropanes in Organic Synthesis

As the proceeding chapters demonstrate, Ni(0)- and Pd(0)-catalyzed [3+2]-cycloadditions of methylenecyclopropanes with alkenes open a new, simple, and useful route to a number of substituted methylenecyclopentanes. This catalytic generation of a "trimethylenemethane" synthon and its addition to olefinic double bonds not only lead to five-membered rings but also introduce an exocyclic methylene group, which is a useful functionality for further structural elaboration.

In addition to the methylenecyclopropane route two other, rather similiar methods exist for the preparation of methylenecyclopentanes, which utilize the transition metal

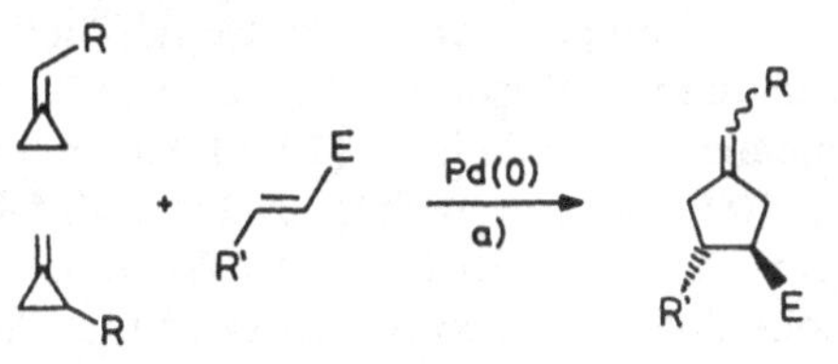

(115a)

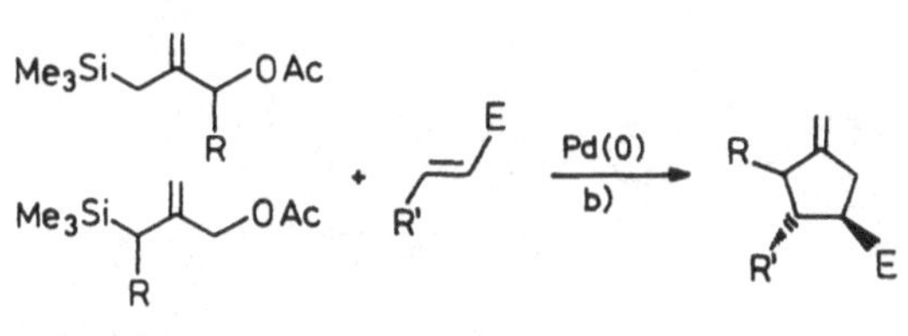

(115b)

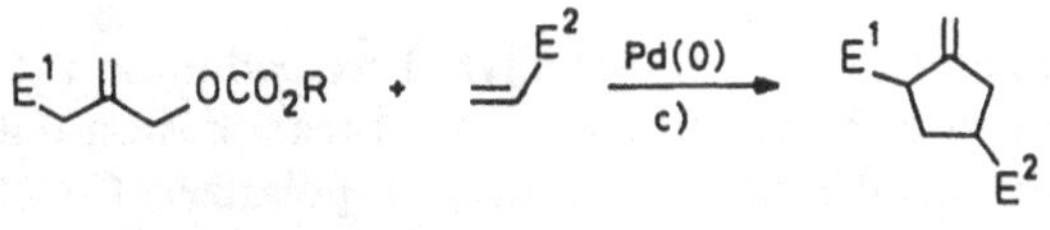

(115c)

catalyzed [3+2]-cycloaddition of a "trimethylenemethane" synthon to electron deficient olefins [183,201,184]. These approaches, mentioned above (see Eq. 77 in Chapter 3.3.1.1) differ from ours in the choice of starting materials used as precursor to the "trimethylenemethane" species; Trost uses (2-(acetoxymethyl)-3-allyl]trimethylsilan [183a] or its derivatives suitably substituted in the 2- or 3-position [183b], whereas Tsuji uses 2-(sulfonylmethyl)- or 2-(cyanomethyl)-3-allyl carbonate [184].

Although these three Pd(0) catalyzed methylenecyclopentane syntheses appear to differ mainly in the choice of starting materials, a closer inspection reveals substantial differences in scope, in the nature of the catalyst, and in mechanism.

Reaction (115a) proceeds with olefins such as norbornene, ethylene or 1-alkenes as well as with electron-deficient olefins. Best results are obtained with alkylphosphanes and a P:Pd ratio of 1:1. Increasing this ratio can totally inhibit the reaction. On the other hand, Reactions 115b and 115c require electron-deficient olefins; they proceed best with a Pd(0) catalyst that bears an arylphosphane or -phosphite ligand and, in general, with a P:Pd ratio of 4:1. With substituted starting materials, quite different regioselectivity is observed in reactions 115a and 115b. While the reaction in Eq. 115a can also be carried out with Ni(0) catalysts (Eq. 116) nothing is known about the use of Ni(0) catalysts in reactions like 115b or 115c. It is likely that they do not work because of the redox stability of nickel acetates or alkoxides. The different regiochemistry observed with Ni(0) catalysts is notable.

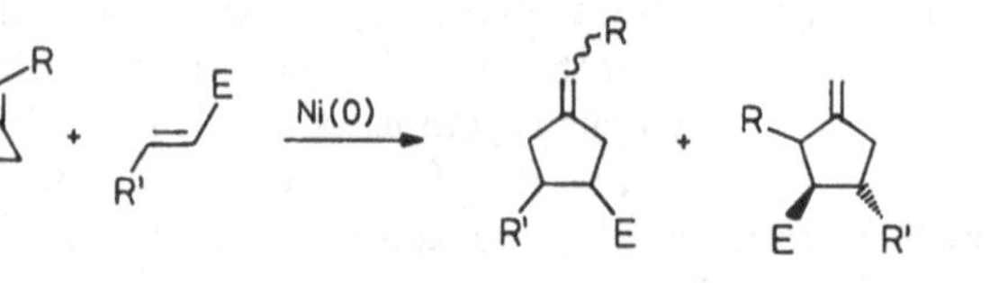

(116)

All these observations indicate that different reaction mechanisms are responsable for the individual characteristics of the related codimerizations summarized in Eqs. 115 and 116.

Theoretical studies [225,226] as well as preparative work strongly indicate that the reactive palladium organic intermediate in Reaction 115b and 115c is an unsymmetrical, zwitterionic trimethylenemethane-palladium (TMM-Pd) complex, as formulated in Eq. 117. Moreover, cycloaddition with a cyclic TMM-Pd-precursor revealed that the electron-deficient olefin attacks the TMM-Pd unit from the side away from the metal. This demonstrates that complexation of the olefin with the metal does not occur prior to C—C bond formation [183].

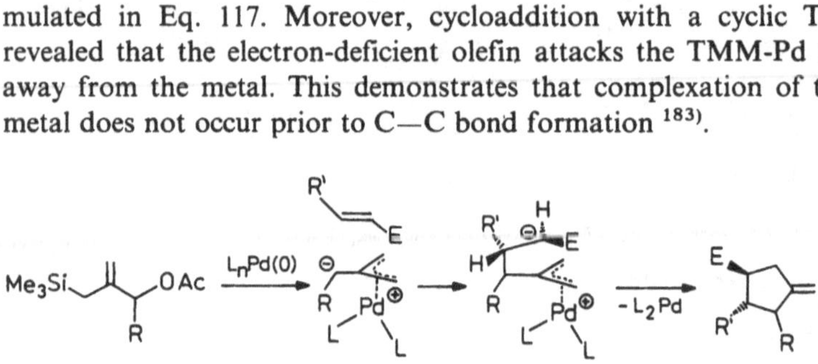

(117)

The situation at the beginning of the metal-catalyzed cycloaddition with methylenecyclopropanes is somewhat different because one has to break the non-polarized distal C—C σ-bond of the three-membered ring and not a polarized C—O-band as in

Eq. 117. Since breaking this C—C bond requires more energy, it is not necessarily so that the most energetically favoured TMM-Pd species act as intermediate, TMM-Pd-complexes with a higher energy level may be reasonable intermediates if they react immediately with a second olefin. We believe that in the cycloaddition reaction with methylenecyclopropane the first metalorganic intermediate is a complex in which both olefins are coordinated to the metal and so held in proximity to each other. Further reaction may involve a direct or a stepwise coupling as shown in Eq. 118. It is unknown at the present time which process occurs and whether it also depends on the nature of the metal and of the second olefin.

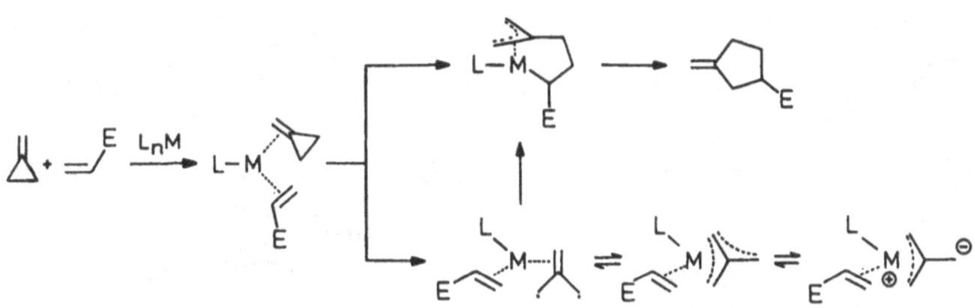

(118)

The following observations are in agreement with these proposals

a) only those olefins react which are known to coordinate well with Pd and Ni;
b) a competitive cyclodimerization and codimerization can be avoided by using methylenecyclopropanes substituted at the double bond which coordinate less readily.

A further support for the mechanism outlined in Eq. 118 is that with Ni(0) catalysts a second type of [3+2]-cycloaddition can occur which involves the oxidative coupling of two alkenes coordinated at the nickel (one must be methylenecyclopropane). The initially formed nickelacyclopentane derivative may collapse to give a spiro[2.3]cyclohexane derivate or rearrange into a 4-methylenenickelacyclohexane derivate, which at the end of this catalytic cycle gives methylenecyclopentanes with a new substitution pattern by reductive elimination (see Eq. 78 and Scheme 8).

The two Pd(0) or Ni(0) catalyzed [3+2]-cycloadditions starting with the readily accessible "trimethylenemethane"-precursors [2-(acetoxymethyl)-3-allyl]trimethylsilan, methylenecyclopropane, and their substituted derivatives are important new methods for the synthesis of methylenecyclopentanes. Because of the simplicity with which many problems of cyclopentane-syntheses can be solved in a convenient one pot reaction this new methodology may be compared with the synthesis of six-membered rings by the powerful [4+2]-cycloaddition of the Diels-Alder reaction.

In many cases [3+2]-cycloaddition with both types of TMM-precursors lead to the same methylenecyclopentanes; but many examples also exist in which the methods give different results, complementary to each other. Some of the most striking differences between the behaviour of Trost's reagents and methylenecyclopropanes with Pd(0) catalysts are summarized in Eqs. 115a and 115b. Two advantages of using methylenecyclopropanes as starting materials are their ability

a) to codimerize not only with electron deficient olefins but also with normal olefins and
b) to codimerize with both Pd(0) and Ni(0) catalysts.

Table 11. Methylenecyclopentanes by Pd(0)/Ni(0)-catalyzed [3+2]-cycloaddition of TMM-precursors and electron deficient olefines

TMS-precursor	Olefine	Catalyst	Condition [°C](h)	Codimere [yield %] (*trans:cis* ratio)
A	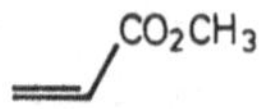	$Pd[PPh_3]_4$	87 (43)	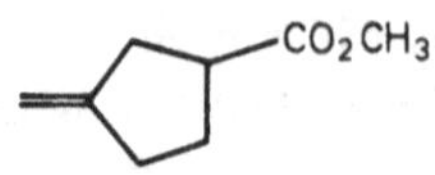 [68] [183a)]
B		$Pd(0)/1\ P(iPr)_3$	130 (3)	[84] [193)]
B		$Ni(cod)_2$	20 (5)	[92] [175)]
A	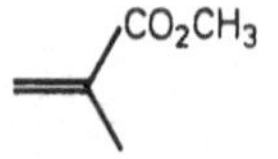	$Pd[PPh_3]_4$	85 (67)	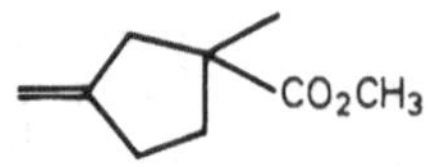 [50] [183)]
B		$Pd(0)/1\ P(iPr)_3$	120 (10)	[< 10] [27)]
A	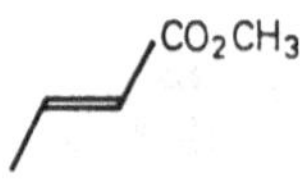	$Pd[PPh_3]_4$	110 (60)	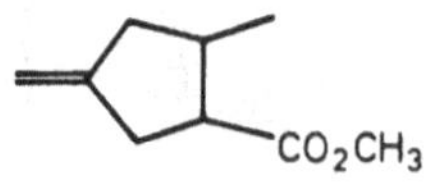 [30] [183a)] (31:1)
B		$Pd(0)/1\ P(iPr)_3$	100 (16)	[43] [193)] (*trans*)
C		"	130 (5)	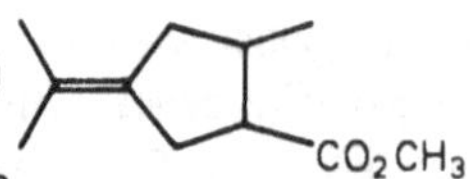 [71] [27)] (*trans*)
D		"	110 (5)	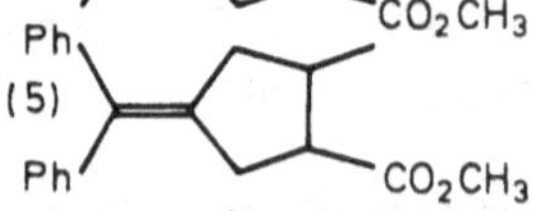 [94] [230)] (*trans*)
B		$Ni(cod)_2$ /1 TOPP	100 (5)	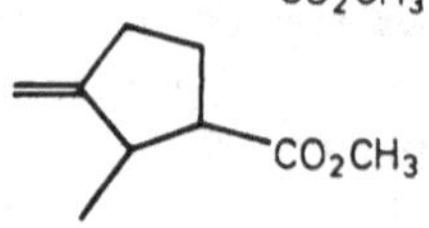 [55] [189)] (78 : 22)
A	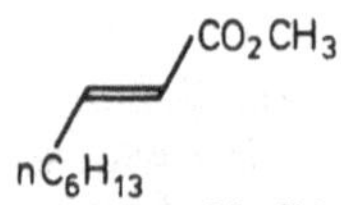	$Pd[PPh_3]_4$	65 (12)	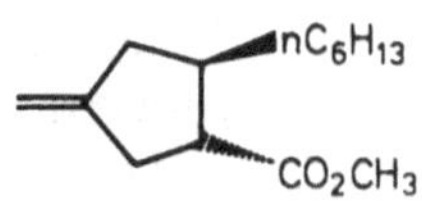 [51] [183a)]
C	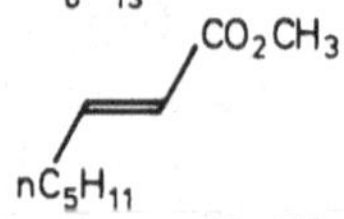	$Pd(0)/1\ P(iPr)_3$	130 (10)	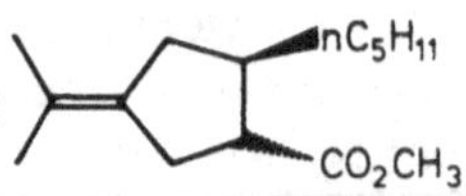 [61] [230)]
D	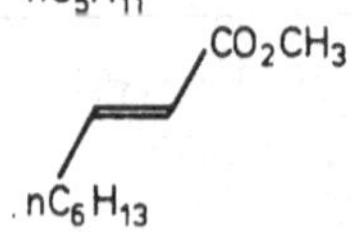	$Pd(0)/1\ P(iPr)_3$	120 (5)	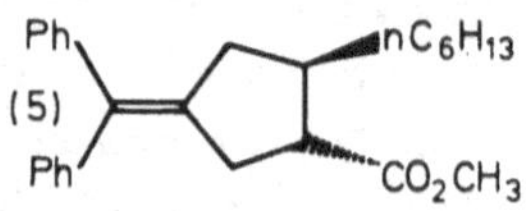 [98] [230)]
A		$Pd[PPh_3]_4$	60 (150)	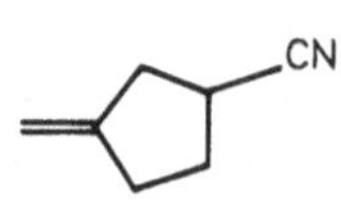 [35] [183a)]
B	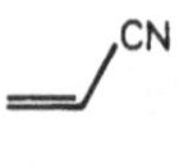	$Pd(0)/1\ P(iPr_3)$	80-130	[0] [a);27)]

Table 11. (continued)

TMS-precursor	Olefine	Catalyst	Condition [°C](h)	Codimere [yield %] (*trans:cis* ratio)	
C		Ni(cod)$_2$/1 TOPP	120 (5)	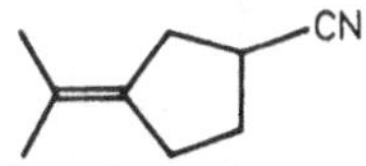	[42] [b);27)]
D		" / "	120 (5)	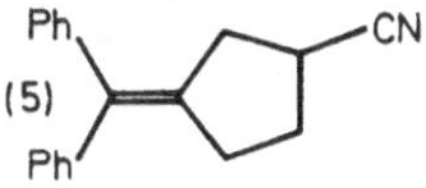	[82] [27)]
A	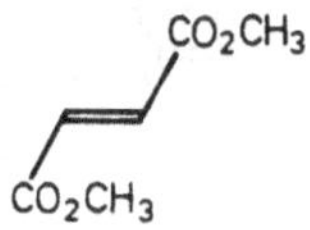	Pd[PPh$_3$]$_4$	65 (285)	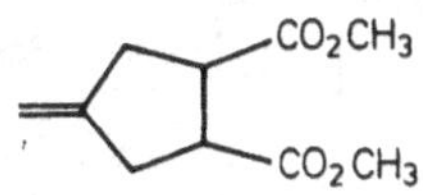	[32] [183a)] (>99:1)
B	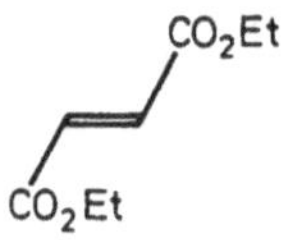	Pd(O)/1 P(iPr)$_3$	120 (2)	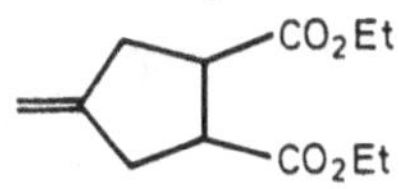	[87] [193)] (99:5)
D		" / "	100 (3)	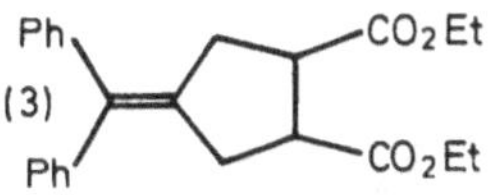	[100] [230)] (*trans*)
A	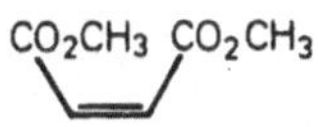	Pd[PPh$_3$]$_4$	65 (210)	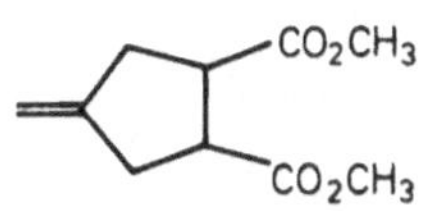	[60] [183a)] (13:1)
B		Pd(O)/1 P(iPr)$_3$	100 (16)		[68] [193)] (43:57)
B		" / "	130 (2)		[64] [c);195)] (2:8)
B	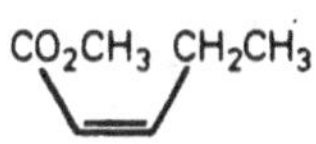	Ni(cod)$_2$	40 (20)	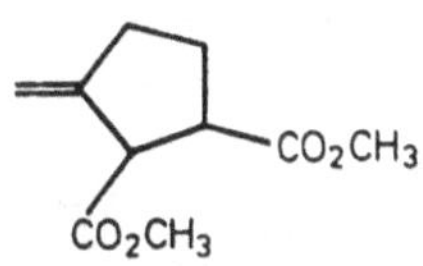	[78] [175)] (1:9)
A	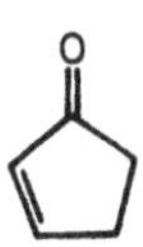	Pd[PPh$_3$]$_4$	65 (20)	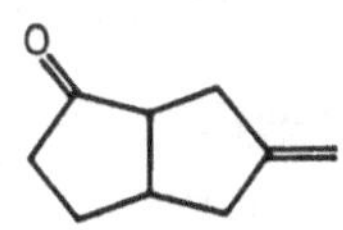	[56] [183)]
B		Pd(O)/1 P(iPr)$_3$	130 (2)		[65] [c);195)]
C		" / "	100 (3)	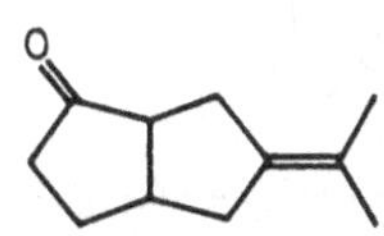	[71] [230)]
D		" / "	90 (3)	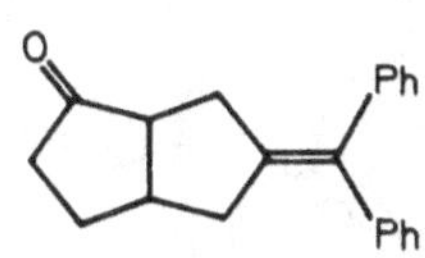	[90] [200)]

Table 11. (continued)

A		Pd[PPh_3]$_4$	65 (20)	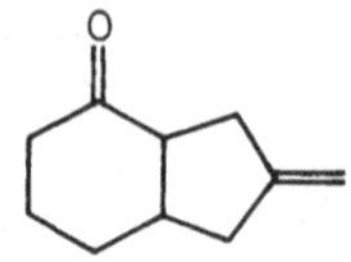	[17] (*cis*) 183)
C	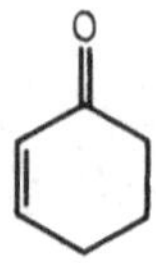	Pd(0)/1 P(iPr)$_3$	130 (10)	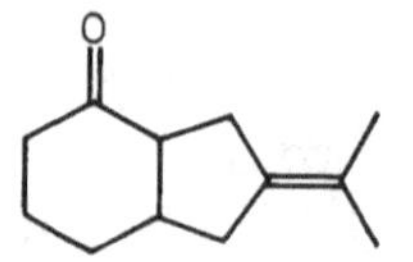	[68] (2:8) 230)
D		" / "	125 (3)	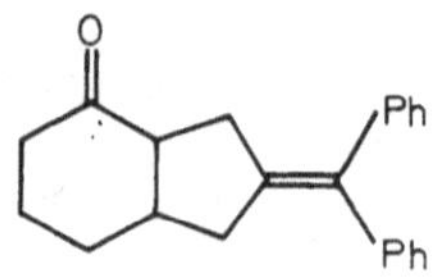	[76] (24:76) 200)
C	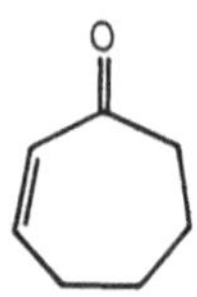	Pd(0)/1 P(iPr)	130 (10)	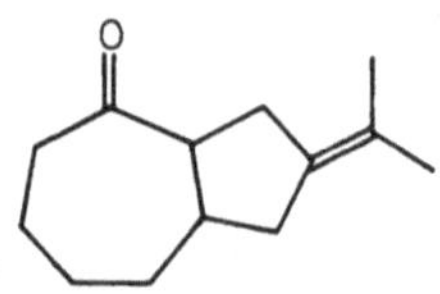	[46] (88:12) d); 230)
D		" / "	125 (7)	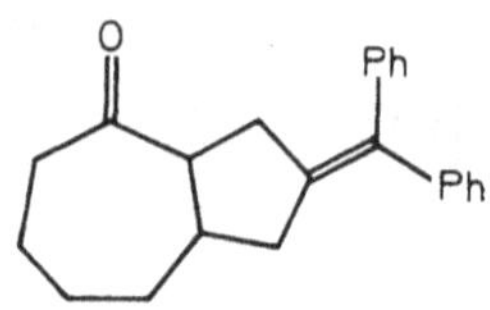	[80] (84:16) 200)

A : 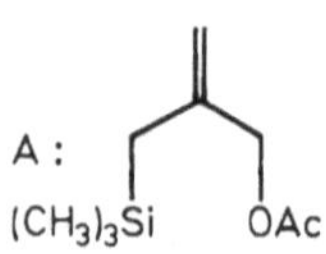

; B :  ; C :  ; D : 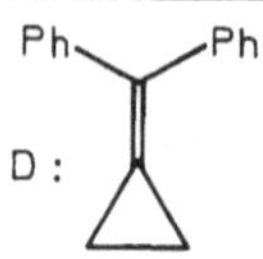

a) No reaction occurs ;

b) 9% 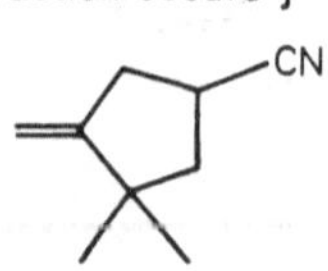

are also formed ;

c) Reactants were pumped into the autoclave heated to 130°C and charged with the catalyst

d) 5% 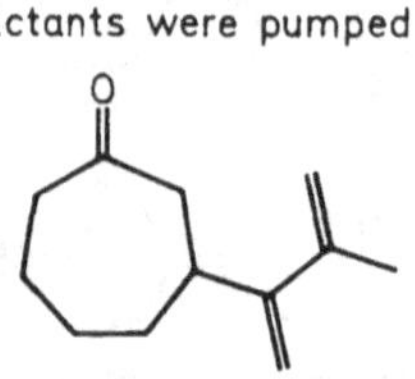 are also formed.

With Ni(0) catalysts they can react by two different reaction mechanism which give the possibility of synthesizing methylene-cyclopentanes with different substituent patterns (see Eqs. 81, 86 and 87). In Table 11 some of the most interesting results obtained in the preparation of methylenecyclopentanes from [(2-acetoxymethyl)-3-allyl]trimethylsilan or methylenecyclopropanes and electron deficient olefins by this new [3+2]-cycloaddition methodology are summarized.

In addition to the above discussed transition metal catalyzed [3+2]-cycloadditions some other methylenecyclopentane syntheses have been developed which are of special interest within the context of annelation procedures.

They are summarized in Eqs. 119a–c.

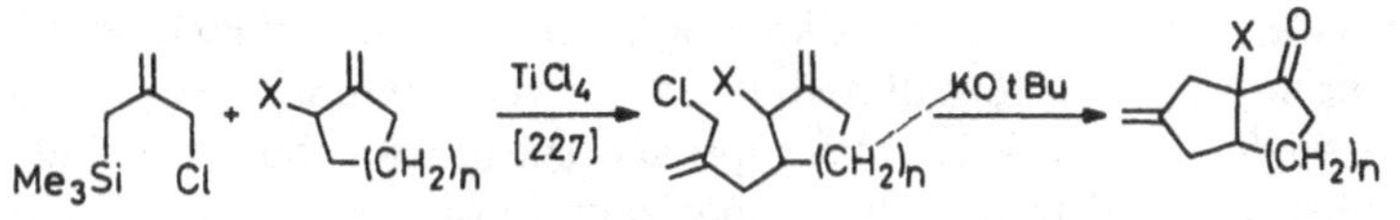

(119a)

n = 1–2 x = H, SPh

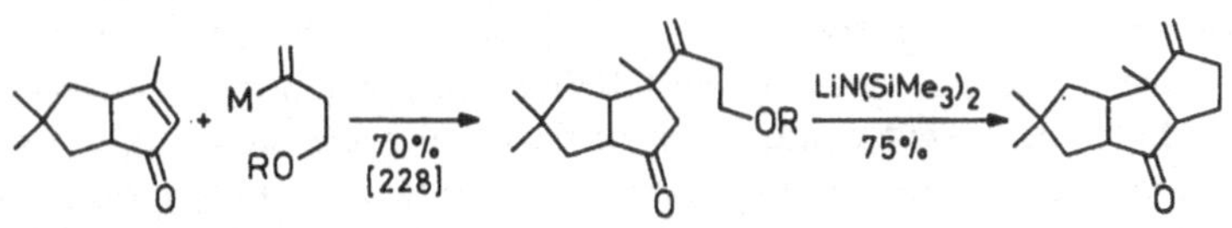

(119b)

M = LiCuJ

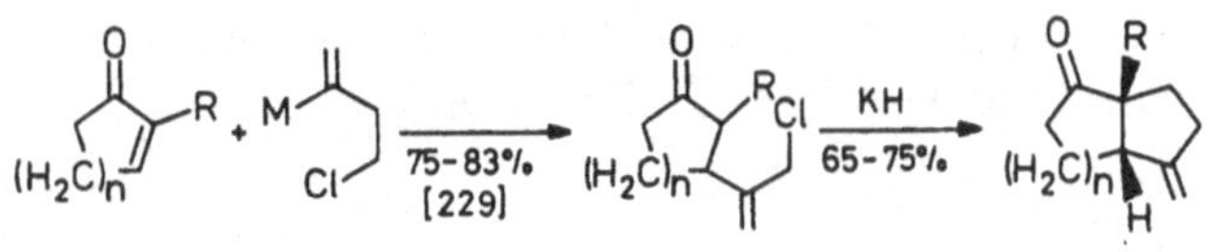

(119c)

M = Cu(SPh)Li n = 1–2 R = H, CH_3
Cu(CN)Li

While reaction 119b has been used for a (+/—) hirsutene synthesis, the annelation according 119c was elaborated for a (+/—) pentalene preparation [229b)].

A common feature of all these methylenecyclopentane syntheses is the incoporation of a methylen group into the new formed five-membered ring, which allows further functionalization. Among the manifold possibilities of derivatisation oxidation to cyclopentanones and cyclopropanation to spiro[4,2]heptanes are of particular importance (Eq. 120). The spiro[4,2]heptanes thus formed can easily be hydrogenated to give cyclopentanes possessing a geminal dimethyl group [231–233)].

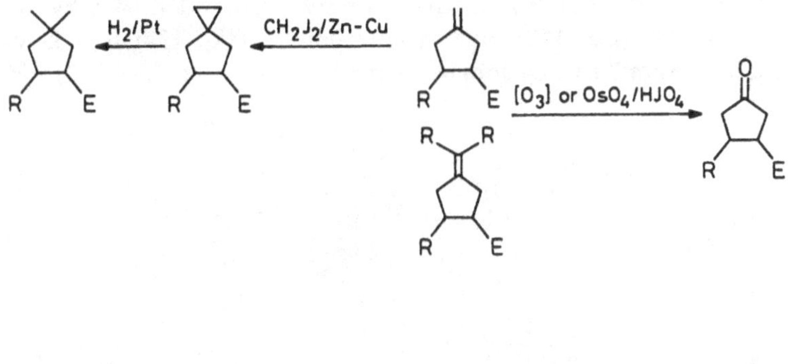

(120)

Methylenecyclopentanes and alkylidencyclopentanes are easily converted by oxidation into the corresponding cyclopentanone derivatives in high yield (90–95%). The oxidations of (diphenylmethylene)cyclopentanes are more difficult: these reactions are often imcomplete or do not work at all.

Starting from a methylenecyclopropane, the two step sequence — catalyzed codimerization to a methylenecyclopentane and subsequent oxidation of the exomethylene group — is often much more convenient for the preparation of specially substituted cyclopentanones than hitherto known procedures. 3-Oxocyclopentane methylcarboxylate, for example, has been prepared in four steps starting with dimethylmalonate and dimethylitaconate (overall yield: 18%) [234], the new two step procedure furnishes this compound in an overall yield of 83%.

Another example is the preparation of 1,2-di(methoxycarbonyl)-4-cyclopentanone. This compound has been prepared from dimethylmaleate as a mixture of its *cis*- and *trans*-isomers (yield: 45–55%) [235]. A second approach starts with the Diels-Alder adduct of 1,3-butadiene and maleic anhydride and leads to the *trans*-diacid (yield: 40%) [236]. The oldest preparation of 1,2-di(methoxycarbonyl)-4-cyclopentanone stems from Auwers in 1893 [237]. Now, the *trans*-isomer is available from methylenecyclopropane and dialkylfumerate in an overall yield of 85%, whereas the pure *cis*-isomer can be obtained from methylenecyclopropane and dialkylmaleate in 53% yield. With dimethylmaleate, the Pd(0) catalyzed codimerization leads to a *cis/trans*-mixture of 4-methylene-1,2-cyclopentane-dimethylcarboxylate in the ratio 8:2 under optimal conditions (see Table 11); after oxidation, the pure *cis*-cyclopentanone derivative can be isolated by simple crystallization from pentane [193,195].

4-Oxo-*trans*-1,2-cyclopentane dimethylcarboxylate is the starting material for two total syntheses of racemic Brefeldine A [236b,238] (Eq. 121).

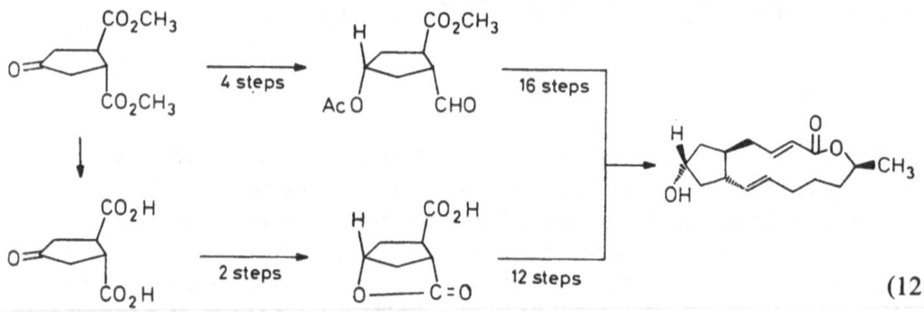

(121)

The same compound has also been used to prepare 11-deoxy-11α-hydroxy-methyl prostaglandin E_s [239].

Beginning with 3-oxycyclopentane methylcarboxylate some natural products have also been synthesized. The key intermediate for the preparation of dihydrojasmone [240], *cis*-jasmone [240], (Eq. 123) and sarkomycin [241] (Eq. 124) is a keto-lactone, which is obtained from 3-oxo-cyclopentane methylcarboxylate (Eq. 122) [239a, 240].

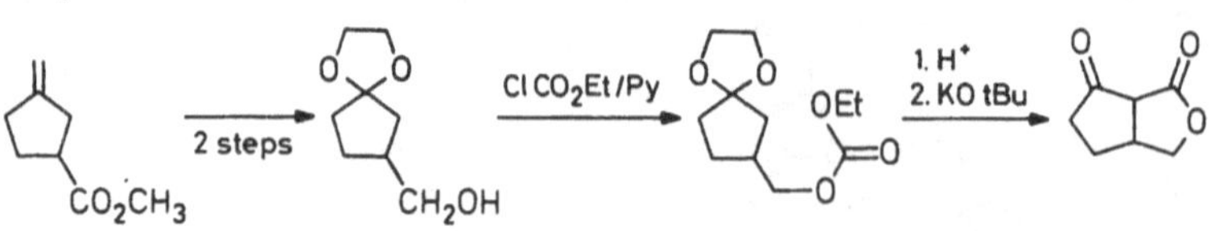

(122)

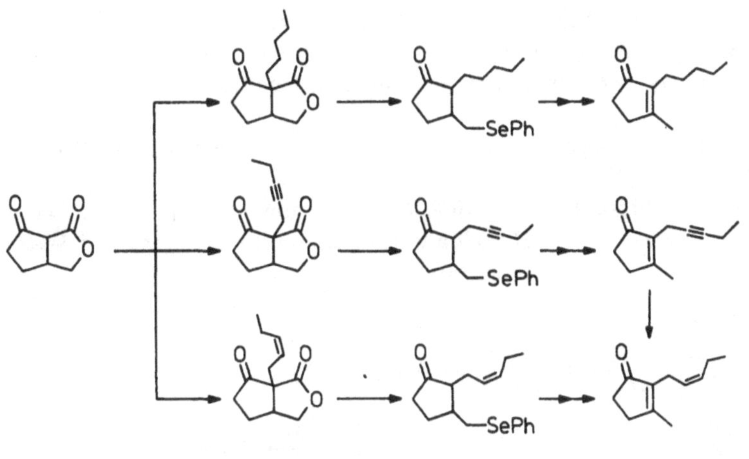

(123)

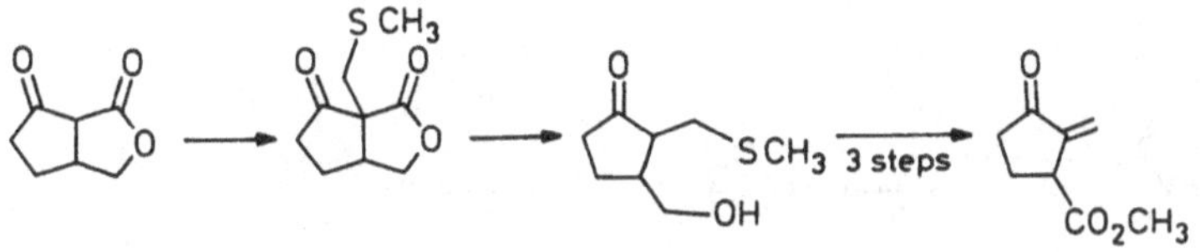

(124)

Instead of the ketolactone, a trimethylsilyl enolate may also serve as starting material for some of the above mentioned syntheses. This can be prepared with TMSCl and DBU as a 6:4 mixture of 3-trimethylsiloxy-2- and -3-cyclopentene-methylcarboxylate, from which the pure lower boiling 2-cyclopentene isomer is obtained by fractional destillation [195] (Eq. 125).

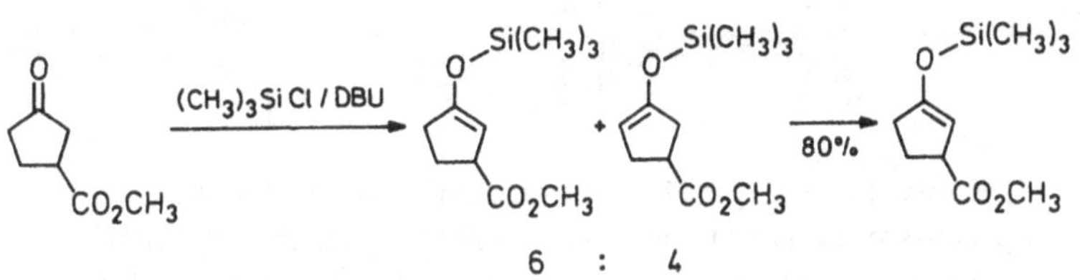

(125)

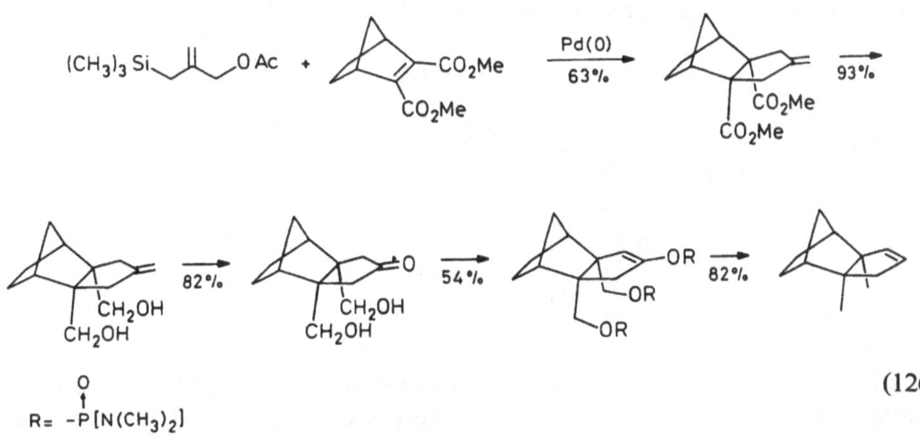

(126)

Other natural product syntheses based on Pd(0)catalyzed [3+2]-cycloadditions have been developed by B. M. Trost using [2-(acetoxymethyl)-3-allyl]trimethylsilane. Thus (+/—) albene has been prepared according to the reaction sequence of Eq. 126 [242].

The initially formed [3+2]-cycloadduct can also be synthesized from methylene-cyclopropane in 68% yield [243].

A loganin aglucon synthesis has been developed as shown in Eq. 127 [245]. Here the key step is the regioselective [3+2]-cycloaddition of 2-cyclopentenone and 3-[(tri-methylsilyl)methyl]-3-buten-2-yl methylcarbonate.

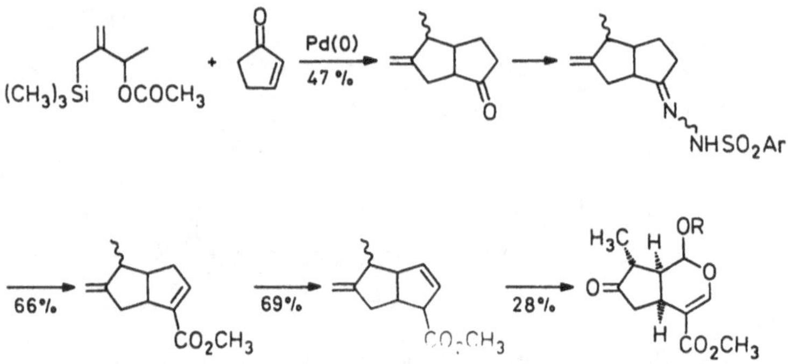

(127)

2-Methyl-3-methylene bicyclo[3.3.0]octan-6-one is also the starting material for a chrysomelidial synthesis (Eq. 128) [245/246].

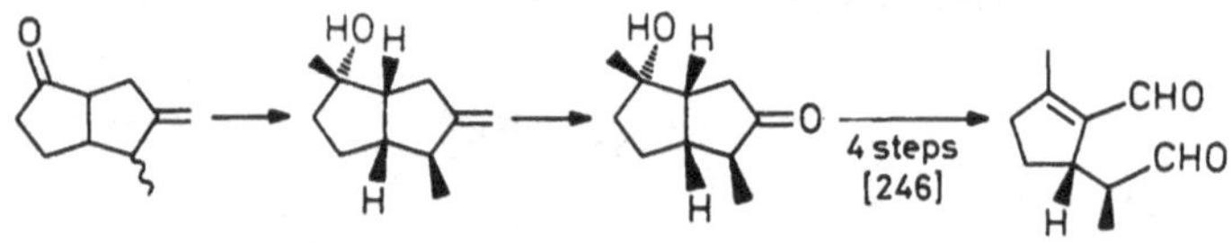

(128)

Recently the Pd(0) catalyzed [3+2]-cycloaddition methodology has been used to prepare a synthetic precursor of pentalenene (Eq. 129) [247]. In this approach, the transformation of the exo-methylene into a geminal dimethyl-group via cyclopropanation has been applied to produce the desired product.

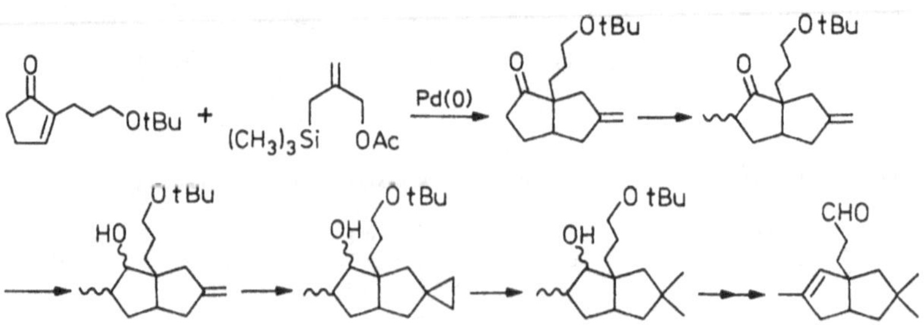

(129)

In summary, the concept of the metal-catalyzed [3+2]-cycloaddition using a tri-methylenemethane-metal intermediate is a very promising method for the direct synthesis of five-membered rings in a one pot procedure. Since the TMM-precursors, especially the methylenecyclopropanes discussed in detail here, are now conveniently available in multigramm quantities this new methodology could well be developed into a five-membered ring synthesis as universally applicable as the Diels-Alder reaction is for six-membered ring synthesis. Further investigations are required to achieve this aim, but the vigorous activity and wide interest in this field assure its rapid development in the near future.

4 Conclusion

Cyclopropene, methylenecyclopropane and their derivatives have proved to be valuable reagents in transition metal-catalyzed cycloaddition reactions. Small and medium carbocycles can be prepared by this method. The chemoselectivity observed in some of these reactions is quite remarkable. In addition, high degree of regio- and stereoselectivity is obtained in most cases. In particular the new [3+2] cycloaddition described here and which involves methylenecyclopropane and its derivatives as trimethylenemethane synthones, shows great synthetic promise as a method for constructing fivemembered rings.

However, there are still a number of problems that will have to be solved: Synthetic equivalents will have to be found for those cosubstrates that do not undergo this cycloaddition. The possibilities of inversion the polarity (e.g. by the introduction of a nitro-group at the unsaturated cosubstrate) remain to be examined as does the extension of these cycloaddition reactions to an intramolecular version leading to cyclopentane annulation. Although the regioselectivity may be manipulated by changing the catalyst (e.g. Pd(0) versus Ni(0)), there is still room for improvement and a deeper understanding of the mechanistic details of these reactions is needed. But the activity and the interest in this field assure that most of these problems will be solved in the near future.

5 Acknowledgements

It is a great pleasure to acknowledge the skilful and enthusiastic contributions of the various friends and coworkers cited in the references. We also thank our colleagues from the analytical departments of the MPI für Kohlenforschung for their valuable assistance. Furthermore, we would like to express our gratitude to the director of the institute, Professor Dr. Dr. h.c. mult. G. Wilke, for his interest as well as for his generous support. H.M.B. is grateful to the Alexander von Humboldt Foundation for the grant of a Feodor-Lynen stipendium and to Professor B. M. Trost and his group for the enjoyable experience of a fruitful stay in Madison.

6 References

1. Köster, R., Arora, S., Binger, P.: Angew. Chem. *81*, 186 (1969); Angew. Chem. Int. Ed. Engl. *8*, 205 (1969)
2. Köster, R., Arora, S., Binger, P.: Synthesis 322 (1971)

3. Salaün, J. R., Conia, J. M.: J. Chem. Soc. Chem. Commun. 1579 (1971)
4. Caubere, P., Coudert, G.: Bull. Soc. Chim. France, 2234 (1971)
5. Köster, R., Arora, S., Binger, P.: Liebigs Ann. Chem. 1219 (1973)
6. Arora, S., Binger, P.: Synthesis 801 (1974)
7. Arora, S., Binger, P., Köster, R.: Synthesis 146 (1973)
8. Binger, P.: Synthesis 190 (1974)
9. Magid, R. M., Clarke, T. C., Duncan, C. D.: J. Org. Chem. *36*, 1320 (1971)
10. Yoshida, Z., Miyahara, H.: Chem. Lett. 335 (1972)
11. Huisgen, R.: Angew. Chem. *75*, 742 (1963); Angew. Chem. Int. Ed. Engl. *2*, 633 (1963)
12. Kellogg, R. M.: Tetrahedron *32*, 2165 (1976)
13. Carter, F. L., Frampton, V. L.: Chem. Rev. *64*, 497 (1964)
14. Closs, G. L.: Adv. Alicyclic Chem. *1*, 53 (1966)
15. Wendisch, D.: Houben/Weyl *4/3*, 679 (1971)
16. Grigorova, T. N.: Sovrem. Probl. Org. Khim. *3*, 100 (1974); CA 82: 154586a
17. Allen, F. H.: Tetrahedron *38*, 645 (1982)
18. Deem, M. L.: Synthesis 675 (1972)
19. Schuchardt, U., dos Santos Filho, P. F.: Cienc. Cult. (Sao Paulo) *30*, 161 (1978); CA 89: 75269r
20. Padwa, A.: Acc. Chem. Res. *12*, 310 (1979)
21. Padwa, A.: Organic Photochem. *4*, 261 (1979)
22. Binger, P., Cetinkaya, B., Doyle, M. J., Germer, A., Schuchardt, U.: Fundam. Res. Homogeneous Catal. *3*, 271 (1979)
23. Deem, M. L.: Synthesis 701 (1982)
24. see ref. 8
25. Nefedov, O. M., Dolgii, I. E., Bulusheva, E. V., Shteinshneider, A. Ya.: Izv. Akad. Nauk SSSR, 1535 (1979); Engl. transl. *28*, 1422 (1979)
26. Dijachenko, A. I., Agre, S. A., Rudashevskaya, T. Y., Shafran, R. M., Nefedov, O. M.: Izv. Akad. Nauk SSSR, 2820 (1984); Engl. transl. *33*, 2585 (1984)
27. Binger, P.: unpublished results
28. Martel, B., Hiriat, J. M.: Synthesis 201 (1972)
29. Demjanov, N. Y., Doyarenko, M. N.: Bull. Acad. Sciences USSR *16*, 297 (1922)
30. Weigert, F. J., Baird, R. L., Shapley, J. R.: J. Am. Chem. Soc. *92*, 6630 (1970)
31. Hoffmann, H. M. R.: Angew. Chem. *81*, 597 (1969); Angew. Chem. Int. Ed. Engl. *8*, 556 (1969)
32. Closs, G. L., Closs, L. E., Böll, W. A.: J. Am. Chem. Soc. *85*, 3796 (1963)
33. v. R. Schleyer, P., Williams, J. E., Blanchard, K. R.: J. Am. Chem. Soc. *92*, 2377 (1970)
34. Bastiansen, O., Fritsch, F. N., Hedberg, K.: Acta Cryst. *17*, 538 (1964)
35. Wiberg, K. B., Ellison, G. B., Wendoloski, J. J., Pratt, W. E., Harmony, M. D.: J. Am. Chem. Soc. *100*, 7837 (1978)
36. Wiberg, K. B., Bartley, W. J.: J. Am. Chem. Soc. *82*, 6375 (1960)
37. Greenberg, A., Liebman, J. F.: Strained Organic Molecules, Academic Press, New York (1978)
38. Davis, J. H., Goddard III, W. A., Bergman, R. G.: J. Am. Chem. Soc. *99*, 2427 (1977)
39. York, E. J., Dittmar, W., Stevenson, J. R., Bergman, R. G.: J. Am. Chem. Soc., *95*, 5680 (1973)
40. de Boer, J. J., Bright, D.: J. Chem. Soc. Dalton Trans. 662 (1975)
41. Binger, P., Doyle, M. J., McMeeking, J., Krüger, C., Tsay, Y.-H.: J. Organomet. Chem. *135*, 405 (1977)
42. Binger, P., Doyle, M. J.: J. Organomet. Chem. *162*, 195 (1978)
43. Peganova, T. A., Petrovskii, P. V., Isaeva, L. S., Kravtsov, D. N., Furman, D. B., Kudryashev, A. V., Ivanov, A. O., Zotova, S. V., Bragin, O. V.: J. Organomet. Chem. *282*, 283 (1985)
44. Binger, P., Büch, H. M., Benn, R., Mynott, R.: Angew. Chem. *94*, 66 (1982); Angew. Chem. Int. Ed. Engl. *21*, 62 (1982); Angew. Chem. Suppl. 153–160 (1982)
45. Cetinkaya, B., Binger, P., Krüger, C.: Chem. Ber. *115*, 3414 (1982)
46. Binger, P., Cetinkaya, B., Krüger, C.: J. Organomet. Chem. *159*, 63 (1978)
47. Newton, M. G., Pantaleo, N. S., King, R. B., Chu, C.-K.: J. Chem. Soc. Chem. Commun. 10 (1979)

48. a) Dettlaf, G., Behrens, U., Weiss, E.: Chem. Ber. *111*, 3019 (1978)
b) Klimes, A., Weiss, E.: Chem. Ber., *115*, 2175 (1982); ibid. *115*, 2606 (1982)
49. Franck-Neumann, M., Dietrich-Buchecker, C., Khemiss, A. K.: J. Organomet. Chem. *220*, 187 (1981)
50. Jens, K. J., Weiss, E.: Chem. Ber., *117*, 2469 (1984)
51. Green, M., Norman, N. C., Orpen, A. G.: J. Organomet. Chem. *221*, C11 (1981)
52. Barker, G. K., Carroll, W. E., Green, M., Welch, A. J.: J. Chem. Soc. Chem. Commun., 1071 (1980)
53. Büch, H. M., Krüger, C.: Acta Cryst. *C40*, 28 (1984)
54. Puddephatt, R. J.: Comments Inorg. Chem. *2*, 69 (1982)
55. a) Wong, W., Singer, S. J., Pitts, W. D., Watkins, S. F., Baddley, W. H.: J. Chem. Soc. Chem. Commun. 672 (1972)
b) Visser, J. P., Ramakers-Blom, J. E.: J. Organomet. Chem. *44*, C63 (1972)
56. Schaverien, C. J., Green, M., Orpen, A. G., Williams, I. D.: J. Chem. Soc. Chem. Commun. 912 (1982)
57. Lehmkuhl, H., Mehler, K.: Liebigs Ann. Chem., 2244 (1982); and cited lit. therein
58. Nesmeyanova, O. A., Kudryatseva, G. A.: Izv. Akad. Nauk SSSR, Ser. Khim., 2629 (1982); Engl. transl. *31*, 2324 (1982)
59. Binger, P., Brinkmann, A.: unpublished results
60. Tolman, C. A.: Chem. Rev. *77*, 313 (1977)
61. Franck-Neumann, M., Dietrich-Buchecker, C.: Tetrahedron Lett. *21*, 671 (1980)
62. Matsui, Y., Orchin, M.: J. Organomet. Chem. *244*, 369 (1983)
63. Kirmse, W.: Carbene Chemistry, Organic Chemistry, Monograph Series *1*, 267 Academic Press New York (1971)
64. a) Stechl, H. H.: Angew. Chem. *75*, 1176 (1963); Angew. Chem. Int. Ed. Engl. *2*, 743 (1963)
b) Stechel, H. H.: Chem. Ber. *97*, 2681 (1964)
65. a) Franck-Neumann, M., Lohmann, J. J.: Angew. Chem. *89*, 331 (1977); Angew. Chem. Int. Ed. Engl. *16*, 323 (1977)
b) Baird, M. S., Buxton, S. R., Whitley, J. S.: Tetrahedron Lett. *25*, 1509 (1984)
c) Baird, M. S.: Tetrahedron Lett. *25*, 4829 (1984)
d) Domnin, I. N., Kortikov, R. R., deMeijere, A.: Zh. Org. Khim. SSSR *19*, 2206 (1982); Engl. transl. *19*, 1921 (1983)
e) Boger, D. L., Brotherton, C. E.: Tetrahedron Lett. *25*, 5611 (1984)
66. a) Binger, P., McMeeking, J.: Angew. Chem. *86*, 518 (1974); Angew. Chem. Int. Ed. Engl. *13*, 466 (1974)
b) Binger, P., McMeeking, J., Schäfer, H.: Chem. Ber. *117*, 1551 (1984)
67. a) Tomilov, Yu. V., Bordakov, V. G., Tsvetkova, N. M., Dolgii, I. E., Nefedov, O. M.: Izv. Akad. Nauk SSSR, Ser. Khim., 2413 (1982); Engl. transl. *31*, 2129 (1982)
b) Tomilov, Yu. V., Bordakov, V. G., Tsvetkova, N. M., Shteinshneider, A. Ya., Dolgii, I. E., Nefedov, O. M.: Izv. Akad. Nauk SSSR, Ser. Khim. 336 (1983); Engl. transl. *32*, 300 (1983)
c) Dolgii, I. E., Tomilov, Yu. V., Shteinshneider, A. Yu., Nefedov, O. M.: Izv. Akad. Nauk Nauk SSSR, Ser. Khim. 700 (1983); Engl. transl. *32*, 638 (1983)
68. Ito, Y., Yonezawa, K., Saegusa, T.: J. Org. Chem. *39*, 1763 (1974)
69. Jolly, P. W., Wilke, G.: The Organic Chemistry of Nickel, Vol. 1, Academic Press, New York (1974)
70. a) Schipperijn, A. J., Lukas, J.: Tetrahedron Lett., 231 (1972)
b) idem, Recl. Trav. Chim. Pays-Bas *92*, 572 (1973)
71. Binger, P., Schäfer, H.: Tetrahedron Lett., 4673 (1975)
72. Binger, P., Schäfer, H.: unpublished results
73. Binger, P., McMeeking, J., Schuchardt, U.: Chem. Ber. *113*, 2372 (1980)
74. Binger, P., Schuchardt, U.: Chem. Ber. *114*, 1649 (1981)
75. Bennett, M. J., Purdham, J. T., Takada, S., Masamune, S.: J. Am. Chem. Soc. *93*, 4063 (1971)
76. Gassman, P. G., Johnson, T. H.: J. Am. Chem. Soc. *98*, 861 (1976)
77. Dolgii, I. E., Tomilov, Yu. V., Tsvetkova, N. M., Bordakov, V. G., Nefedov, O. M.: Izv. Akad. Nauk SSSR, Ser. Khim. 958 (1983); Engl. transl. *32*, 868 (1983)

78. Weiß, R., Schlierf, C.: Angew. Chem. *83*, 887 (1971); Angew. Chem. Int. Ed. Engl. *10*, 811 (1971)
79. a) Landheer, I. J., de Wolf, W. H., Bickelhaupt, F.: Tetrahedron Lett., 2813 (1974)
b) idem: ibid. 349 (1975)
80. a) Gröger, C., Musso, H., Roßnagel, I.: Chem. Ber. *113*, 3621 (1980)
b) Stahl, K.-J., Hertzsch, W., Musso, H.: Liebigs Ann. Chem. *1985*, 1474
81. Boger, D. L., Brotherton, C. E.: J. Am. Chem. Soc. *106*, 805 (1984)
82. Komendantov, M. I., D'yakonov, I. A., Smirnova, T. S.: Zh. Org. Khim. *2*, 559 (1966); J. Org. Chem. USSR *2*, 562 (1966)
83. Komendantov, M. I., Kryuchkova, I. K., Domnin, I. N.: Zh. Org. Khim. *6*, 631 (1970); J. Org. Chem. USSR *6*, 630 (1970)
84. Komendantov, M. I., Smirnova, T. S., Domnin, I. N., Krakhmal'naya, L. A.: Zh. Org. Khim. *7*, 2455 (1971); J. Org. Chem. USSR *7*, 2551 (1971)
85. Komendantov, M. I., Domnin, I. N., Bulucheva, E. V.: Tetrahedron *31*, 2495 (1975)
86. Komendantov, M. I., Domnin, I. N.: Zh. Org. Khim. *5*, 1319 (1969); J. Org. Chem. *5*, 1288 (1969)
87. Breslow, R.: Mol. Rearrangements *1*, 257 Interscience Publishers, New York (1963)
88. Battiste, M. A., Halton, B., Grubbs, R. H.: J. Chem. Soc. Chem. Commun. 907 (1967)
89. Walker, J. A., Orchin, M.: ibid. 1239 (1968)
90. Padwa, A., Blacklock, T. J., Loza, R.: J. Am. Chem. Soc. *103*, 2404 (1981)
91. Binger, P., Brinkmann, A.: Chem. Ber. *111*, 2689 (1978)
92. Colquhoun, H. M., Holton, J., Thompson, D. J., Twigg, M. V.: New Pathways for Organic Synthesis, Plenum Press, New York (1984)
93. Woodward, R. B., Hoffmann, R.: Angew. Chem. *81*, 797 (1969); Angew. Chem. Int. Ed. Engl. *8*, 781 (1969)
94. Benson, S. W.: Thermodynamical Kinetics, Wiley, New York (1976)
95. a) Blomquist, A. T., Meinwald, Y. C.: J. Am. Chem. Soc. *81*, 667 (1959)
b) Hall jr., H. K.: J. Org. Chem. *25*, 42 (1960)
c) Williams, J. K., Benson, R. E.: J. Am. Chem. Soc. *84*, 1257 (1962)
d) Cookson, R. C., Gilani, S. S. H., Stevens, I. D. R.: Tetrahedron Lett. 615 (1962)
e) Cookson, R. C., Dance, J., Hudec, J.: J. Chem. Soc. 5416 (1964)
f) Heaney, H., Jablonski, J. M.: Tetrahedron Lett. 2733 (1967)
96. Reppe, W., Schlichting, O., Klager, K., Toepel, T.: Liebigs Ann. Chem. *560*, 1 (1948)
97. Vollhardt, K. P. C.: Angew. Chem. *96*, 525 (1984)
98. Bönnemann, H.: Angew. Chem. *90*, 517 (1978): Angew. Chem. Int. Ed. Engl. *17*, 505 (1978)
99. Binger, P., Schroth, G., McMeeking, J.: Angew. Chem. *86*, 518 (1974); Angew. Chem. Int. Ed. Engl. *13*, 465 (1974)
100. Krüger, C., Roberts, P. J.: Cryst. Struct. Comm. *3*, 459 (1974)
101. Binger, P., McMeeking, J.: Angew. Chem. *87*, 383 (1975); Angew. Chem. Int. Ed. Engl. *14*, 371 (1975)
102. Spielmann, W., Kaufmann, D., de Meijere, A.: Angew. Chem. *90*, 470 (1978); Angew. Chem. Int. Ed. Engl. *17*, 440 (1978)
103. a) Reinhoudt, D. N., Smael, P., van Tilborg, W. J. M., Visser, J. P.: Tetrahedron Lett. 3755 (1973)
b) van Tilborg, W. J. M., Smael, P., Visser, J. P., Kouwenhoven, C. G., Reinhoudt, D. N.: Recl. Trav. Chim. Pays-Bas *94*, 85 (1975)
104. Noyori, R., Umeda, I., Takya, H.: Tetrahedron Lett. 1189 (1972)
105. Binger, P., Schuchardt, U.: Angew. Chem. *87*, 715 (1975); Angew. Chem. Int. Ed. Engl. *14*, 706 (1975)
106. Older methods are cited in: Rahman, W., Kuivila, H. G.: J. Org. Chem. *31*, 772 (1966)
107. Fisher, F., Applequist, D. E.: J. Org. Chem. *30*, 2089 (1965)
108. Rule, M., Matlin, A. R., Hilinski, E. F., Dougherty, D. A., Berson, J. A.: J. Am. chem. Soc. *101*, 5098 (1979)
109. Billups, W. E., Shields, T. C., Chow, W. Y., Deno, N. C.: J. Org. Chem. *37*, 3676 (1972)
110. Billups, W. E., Chow, W. Y., Leavell, K. H., Lewis, E. S.: J. Org. Chem. *39*, 274 (1974)
111. Osborn, C. L., Shields, T. C., Shoulders, B. A., Krause, J. F., Cortez, H. V., Gardner, P. D.: J. Am. Chem. Soc. *87*, 3158 (1965)
112. Bässler, T.: Dissertation, Universität Saarbrücken (1972)

113. Makosza, M., Wawrzyniewicz, M.: Tetrahedron Lett. 4659 (1969)
114. a) Closs, G. L., Closs, L. E.: J. Am. Chem. Soc. *82*, 5723 (1960)
b) see ref. 28
115. Thomas, E. W.: Tetrahedron Lett. *24*, 1467 (1983)
116. Sternberg, E., Binger, P.: Tetrahedron Lett. *26*, 301 (1985)
117. Wittig, U., Binger, P.: unpublished results
118. Schöllkopf, U.: Houben/Weyl *13/1*, 167 (1970)
119. Dunkelblum, E., Brenner, S.: Tetrahedron Lett. 669 (1973)
120. a) Sisido, K., Utimoto, K.: Tetrahedron Lett. 3267 (1966)
b) Utimoto, K., Tamura, M., Sisido, K.: Tetrahedron *29*, 1169 (1973)
c) Schweizer, E. E., Berninger, C. J., Thompson. J. G.: J. Org. Chem. *33*, 336 (1968)
d) Trost, B. M., LaRochelle, R., Bogdanowicz, M. J.: Tetrahedron Lett. 3449 (1970)
e) Huet, F., Lechevallier, A., Conia, J. M.: Tetrahedron Lett. 2521 (1977)
f) Lechevallier, A., Huet, F., Conia, J. M.: Tetrahedron *39*, 3307 (1983)
121. Osborne, N. F.: J. Chem. Soc. Perkin Trans. I, 1435 (1982)
122. a) Hiyama, T., Kanakura, A., Morizawa, Y., Nozaki, H.: Tetrahedron Lett. *23*, 1279 (1982)
b) Hässig, R., Siegel, H., Seebach, D.: Chem. Ber. *115*, 1990 (1982)
c) Halazy, S., Dumont, W., Krief, A.: Tetrahedron Lett. *22*, 4737 (1981)
d) Cohen, T., Sherbine, J. P., Matz, J. R., Hutchins, R. R., McHenry, B. M., Willey, P. R.: J. Am. Chem. Soc. *106*, 3245 (1984)
123. a) see ref. 106
b) Battioni, P., Vo-Quang, L., Vo-Quang, Y.: Bull. Soc. Chim. France 3942 (1970)
c) Dehmlow, E. V.: Chem. Ber. *100*, 2779 (1967)
124. a) Noyori, R., Takaya, H., Nakanisi, Y., Nozaki, H.: Can. J. Chem. *47*, 1242 (1969)
b) Molchanov, A. P., Noskova, O. V., Kostikov, R. R.: Zh. Org. Khim. *19*, 1981 (1983); J. Org. Chem. USSR *19*, 1740 (1983)
c) Creary, X.: J. Org. Chem. *43*, 1777 (1978)
d) Blomquist, A. T., Conolly, D. J.: Chem. Ind. 310 (1962)
125. a) Stang, P. J.: Acc. Chem. Res. *15*, 348 (1982)
b) Fox, D. P., Bjork, J. A., Stang, P. J.: J. Org. Chem. *48*, 3994 (1983)
126. Vogel, E.: Angew. Chem. *72*, 4 (1960)
127. a) Berson, J. A. et al.: J. Am. Chem. Soc. *104*, 2233 (1982); ibid. *104*, 2228 (1982); ibid. *104*, 2223 (1982); ibid. *104*, 2217 (1982); ibid. *104*, 2209 (1982); ibid. *102*, 3870 (1980)
b) Dixon, D. A., Foster, R., Halgren, T. A., Lipscomb, W. N.: J. Am. Chem. Soc. *100*, 1359 (1978)
c) Hehre, W. J., Salem, L., Willcott, M. R.: J. Am. Chem. Soc. *96*, 4328 (1974)
d) Dewar, M. J. S., Wasson, J. S.: J. Am. Chem. Soc. *93*, 3081 (1971)
128. a) Aue, D. H., Lorens, R. B., Helwig, G. S.: J. Org. Chem. *44*, 1202 (1979)
b) see ref. 22
c) Hartmann, W., Heine, H.-G., Hinz, J., Wendisch, D.: Chem. Ber. *110*, 2986 (1977)
d) Noyori, R., Hayashi, N., Kato, M.: Tetrahedron Lett. 2983 (1973)
e) Noyori, R., Ishigami, T., Hayashi, N., Takaya, H.: J. Am. Chem. Soc. *95*, 1674 (1973)
f) Noyori, R., Kumagai, Y., Umeda, I., Takaya, H.: J. Am. Chem. Soc. *94*, 4018 (1972)
g) Barton, T. J., Rogido, R. J.: Tetrahedron Lett. 3901 (1972)
129. Laurie, V. W., Stigliani, W. M.: J. Am. Chem. Soc. *92*, 1485 (1970)
130. Krüger, C., Chiang, A.-P.: unpublished results
131. a) v.d. Saal, W., Risler, W., Stawitz, J., Quast, H.: J. Org. Chem. *48*, 2374 (1983)
b) Barfield, M., Spear, R. J., Sternhell, S.: Chem. Rev. *76*, 593 (1976)
c) Rol, N. C., Clague, A. D. H.: Org. Magn. Reson. *16*, 187 (1981)
d) Crawford, R. J., Tokunaga, H., Shrijver, L. M. H. C., Godard, J. C., Nakashima, T.: Can. J. Chem. *56*, 992 + 998 (1978)
e) Aue, D. H., Meshishnek, M. J.: J. Am. Chem. Soc. *99*, 223 (1977)
f) Günther, H., Herrig, W.: Chem. Ber. *106*, 3938 (1973)
132. a) Bertie, J. E., Norton, M. G.: Can. J. Chem. *48*, 3889 (1970); ibid. *49*, 2229 (1971)
b) Bertie, J. E., Sunder, S.: ibid. *50*, 765 (1972)
c) Mitchell, R. W., Merritt, J. A.: Spectrochim. Acta *27a*, 1609 (1971)

133. Sieiro, C., Ponce, C.: Ann. Quim. *72*, 201 (1976)
134. a) Bieri, G., Burger, F., Heilbronner, E., Maier, J. P.: Helv. Chim. Acta *60*, 2213 (1977)
b) Wiberg, K. B., Ellison, G. B., Wendoloski, J. J., Brundle, C. R., Kuebler, N. A.: J. Am. Chem. Soc. *98*, 7179 (1976)
135. Wiberg, K. B. et al.: J. Am. Chem. Soc. *105*, 3638 (1983); ibid. *104*, 5239 (1982); ibid. *104*, 2056 (1982)
136. Stirling, C. J. M.: Pure and Appl. Chem. *56*, 1781 (1984)
137. for a review see: Gajewski, J. J.: Hydrocarbon Thermal Rearrangements, Academic Press, New York (1981)
138. a) Chesick, J. P.: J. Am. Chem. Soc. *85*, 2720 (1963)
b) Gajewski, J. J.: ibid. *90*, 7178 (1968)
c) Gilbert, J. C., Butler, J. R.: ibid. *92*, 2168 (1970)
d) von E. Doering, W., Roth, H. D.: Tetrahedron *26*, 2825 (1970)
e) Dewar, M. J. S., Wasson, J. S.: J. Am. Chem. Soc. *93*, 3081 (1971)
139. a) Yarkony, D. R., Schaefer III, H. F.: J. Am. Chem. Soc. *96*, 3754 (1974)
b) Borden, W. T.: ibid. *97*, 2906 (1975)
c) Davis, J. H., Goddard III, W. A.: ibid. *98*, 303 (1976)
140. Dowd, P., Chow, M.: Tetrahedron *38*, 799 (1982)
141. a) Salaün, J. R., Conia, J. M.: J. Chem. Soc. Chem. Commun. 1579 (1971)
b) Crandall, J. K., Conover, W. W.: J. Org. Chem. *43*, 3533 (1978); here further literature
142. Köster, R., Arora, S., Binger, P.: Liebigs Ann. Chem. 1619 (1973)
143. Binger, P.: Angew. Chem. *84*, 483 (1972); Angew. Chem. Int. Ed. Engl. *11*, 433 (1972)
144. Bartlett, P. D., Wheland, R. C.: J. Am. Chem. Soc. *94*, 2145 (1972)
145. Noyori, R., Hayashi, N., Kato, M.: J. Am. Chem. Soc. *93*, 4948 (1971)
146. Kaufmann, D., de Meijere, A.: Angew. Chem. *85*, 151 (1973); Angew. Chem. Int. Ed. Engl. *12*, 159 (1973)
147. M = Fe: a) Whitesides, T. H., Slaven, R. W.: J. Organomet. Chem. *67*, 99 (1974)
b) Whitesides, T. H., Slaven, R. W., Calabrese, J. C.: Inorg. Chem. *13*, 1895 (1974)
c) Green, M., Howard, J. A. K., Hughes, R. P., Kellett, S. C., Woodward, P.: J. Chem. Soc. Dalton Trans. 2007 (1975)
d) Pinhas, A. R., Samuelson, A. G., Risemberg, R., Arnold, E. V., Clardy, J., Carpenter, B. K.: J. Am. Chem. Soc. *103*, 1668 (1981)
148. M = Ni: a) Englert, M., Jolly, P. W., Wilke, G.: Angew. Chem. *83*, 84 (1971); Angew. Chem. Int. Ed. Engl. *10*, 77 (1971)
b) Binger, P., Doyle, M. J., Benn, R.: Chem. Ber. *116*, 1 (1983)
c) Isaeva, L. S., Peganova, T. A., Petrovskii, P. V., Furmann, D. B., Zotova, S. V., Kudryashev, A. V., Bragin, O. V.: J. Organomet. Chem. *258*, 367 (1983)
149. M = Pd: a) Green, M., Hughes, R. P.: J. Chem. Soc. Dalton Trans. 1880 (1976)
b) Büch, H. M.: Dissertation, Universität Kaiserslautern (1982)
150. M = Fe: see ref. 147a, 147d
151. M = Ni: see ref. 148a
152. M = Pd: see ref. 149a
153. M = Fe: Chisnall, B. M., Green, M., Hughes, R. P., Welch, A. J.: J. Chem. Soc. Dalton Trans. 1899 (1976)
b) see ref. 147d
154. a) M = Fe: 1) Noyori, R., Nishimura, T., Takaya, H.: Chem. Commun. 89 (1969)
2) see ref. 147a
3) Pinhas, A. R., Carpenter, B. K.: J. Chem. Soc. Chem. Commun. 17 (1980); ibid. 15 (1980)
b) M = Mo: 1) Barnes, S. G., Green, M.: J. Chem. Soc. Chem. Commun. 267 (1980)
2) Allen, S. R., Barnes, S. G., Green, M., Moran, G., Trollope, L., Murrall, N. W., Welch, A. J., Sharaiha, D. M.: J. Chem. Soc. Dalton Trans. 1157 (1984)
155. M = Pd: a) Noyori, R., Takaya, H.: Chem. Commun. 525 (1969)
b) Albright, T. A., Clemens, P. A., Hughes, R. P., Hunton, D. E., Margerum, L. D.: J. Am. Chem. Soc. *104*, 5369 (1982)
156. a) M = Rh; Ir: see ref. 147c

b) M = Co: Binger, P., Martin, T. R., Benn, R., Rufinska, A., Schroth, G.: Z. Naturforsch. *39b*, 993 (1984)
157. M = Co: see ref. 156b
158. M = Ni: a) see ref. 148b
b) see ref. 149b
159. Seebach, D.: Houben/Weyl *4/4*, 1 (1971)
160. Corey, E. J., Mitra, R. B., Uda, H.: J. Am. Chem. Soc. *85*, 362 (1963); ibid. *86*, 485 (1964)
161. see ref. 92
162. a) Baldwin, S. W.: Organic Photochem. *5*, 123 (1981)
b) Chapman, O. L., Weiss, D. S.: ibid. *3*, 197 (1973)
c) de Mayo, P.: Acc. Chem. Res. *4*, 41 (1971)
d) Bauslaugh, P. G.: Synthesis 287 (1970)
e) Eaton, P. E.: Acc. Chem. Res. *1*, 50 (1968)
f) Chapman, O. L., Lenz, G.: Organic Photochem. *1*, 283 (1967)
163. a) Mango, F. D., Schachtschneider, J. H.: J. Am. Chem. Soc. *89*, 2484 (1967)
b) Mango, F. D.: Coord. Chem. Rev. *15*, 109 (1975)
164. Heimbach, P.: Angew. Chem. *82*, 550 (1970); Angew. Chem. Int. Ed. Engl. *9*, 528 (1970)
165. Bishop III, K. C.: Chem. Rev. *76*, 461 (1976)
166. Wilke, G.: J. Organomet. Chem. *200*, 349 (1980)
167. Puddephatt, R. J.: Comments Inorg. Chem. *2*, 69 (1982)
168. Stockis, A., Hoffmann, R.: J. Am. Chem. Soc. *102*, 2952 (1980)
169. McKinney, R. J., Thorn, D. L., Hoffmann, R., Stockis, A.: J. Am. Chem. Soc. *103*, 2595 (1981)
170. Binger, P.: Angew. Chem. *84*, 352 (1972); Angew. Chem. Int. Ed. Engl. *11*, 309 (1972)
171. Binger, P.: Synthesis 427 (1973)
172. Binger, P.: unpublished results, see Habilitationsschrift, Universität Kaiserslautern (1982)
173. Noyori, R., Ishigami, T., Hayashi, N., Takaya, H.: J. Am. Chem. Soc. *95*, 1674 (1973)
174. Kaufmann, D., de Meijere, A.: Chem. Ber. *117*, 3134 (1984)
175. Binger, P., Brinkmann, A., Wedemann, P.: Chem. Ber. *116*, 2920 (1983)
176. a) Wassermann, A.: Diels-Alder Reactions, Elsevier, New York (1965)
b) Wollweber, H.: Houben/Weyl *5/Ic*, 977 (1970); Diels-Alder Reaktion, Thieme-Verlag, Stuttgart (1972)
c) Sauer, J., Sustmann, R.: Angew. Chem. *92*, 773 (1980); Angew. Chem. Int. Ed. Engl. *19*, 779 (1980)
e) Petrzilka, M., Grayson, J. I.: Synthesis 753 (1981)
177. a) House, H. O.: Modern Synthetic Reactions, 2nd ed., p. 606–611, 621–623, Benjamin, New York (1972)
b) Jung, M. E.: Tetrahedron *32*, 3 (1976)
c) Gawley, R. E.: Synthesis 777 (1976)
178. a) Paquette, L. A. et al.: Recent Developments in Polycyclopentanoid Chemistry, Tetrahedron Symposia-in-Print *37*, 4357 (1981)
b) Trost, B. M.: Cyclopentanoids: A Challenge for New Methodology, Chem. Soc. Rev. *11*, 141 (1982)
c) Harre, M., Raddatz, P., Walenta, R., Winterfeldt. E.: Angew. Chem. *94*, 496 (1982); Angew. Chem. Int. Ed. Engl. *21*, 480 (1982)
d) Ramaiah, M.: Synthesis 529 (1984)
e) Paquette, L. A.: Top. Curr. Chem. *119*, 1 (1984)
179. a) for an excellent review see: Tietze, L.-F.: Angew. Chem. *95*, 840 (1983); Angew. Chem. Int. Ed. Engl. *22*, 828 (1983)
b) for a recent synthesis of loganin aglucon via cycloaddition methodology see: Trost, B. M., Nanninga, T. N.: J. Am. Chem. Soc. *107*, 1293 (1985)
180. Paquette, L. A., Ternansky, R. J., Balogh, D. W., Kentgen, G.: J. Am. Chem. Soc. *105*, 5446 (1983)
181. a) Dowd, P.: Acc. Chem. Res. *5*, 242 (1972)
b) Berson, J. A.: ibid. *11*, 446 (1978)
182. some specific intramolecular examples are known, e.g.: Little, R. D., Muller, G. W., Venegas, M. G., Caroll, G. L., Bukhari, A., Patton, L., Stone, K.: Tetrahedron *37*, 4371 (1981)

183. a) Trost, B. M., Chan, D. M. T.: J. Am. Chem. Soc. *105*, 2315 and 2326 (1983)
b) Trost, B. M., Nanninga, T. N., Satoh, T.: J. Am. Chem. Soc. *107*, 721 (1985)
c) Trost, B. M., Nanninga, T. N.: ibid. *107*, 1075 (1985)
184. Shimizu, I., Ohashi, Y., Tsuji, J.: Tetrahedron Lett. *25*, 5183 (1984)
185. Kiji, J., Masui, K., Furukawa, J.: Bull. Chem. Soc. Jap. *44*, 1956 (1971)
186. Noyori, R., Odagi, T., Takaya, H.: J. Am. Chem. Soc. *92*, 5780 (1970)
187. a) Noyori, R., Kumagai, Y., Umeda, I., Takaya, H.: J. Am. Chem. Soc. *94*, 4018 (1972)
b) Noyori, R., Yamakawa, M., Takaya, H.: Tetrahedron Lett. 4823 (1978)
188. Binger, P., Brinkmann, A., Richter, W. J.: Tetrahedron Lett. *24*, 3599 (1983)
189. Binger, P., Wedemann, P.: Tetrahedron Lett. *26*, 1045 (1985)
190. Ukai, T., Kawazura, H., Ishii, Y., Bonnet, J. J., Ibers, J. A.: J. Organomet. Chem. *65*, 253 (1974)
191. Tatsuno, Y., Yoshida, T., Otsuka, S.: Inorg. Synth. *19*, 220 (1979)
192. Kühn, A., Werner, H.: J. Organomet. Chem. *179*, 421 (1979)
193. Binger, P., Schuchardt, U.: Chem. Ber. *114*, 3313 (1981)
194. Chen, S., Binger, P.: unpublished results
195. Binger, P., Brinkmann, A.: unpublished results
196. Balavoine, G., Eskenazi, C., Guillemot, M.: J. Chem. Soc. Chem. Commun. 1109 (1979)
197. Trost, B. M., Renaut, P.: J. Am. Chem. Soc. *104*, 6668 (1982)
198. Binger, P., Schuchardt, U.: Chem. Ber. *113*, 3334 (1980)
199. Binger, P., Schuchardt, U.: Chem. Ber. *113*, 1063 (1980)
200. Binger, P., Bentz, P.: Angew. Chem. *94*, 636 (1982); Angew. Chem. Int. Ed. Engl. *21*, 622 (1982); Angew. Chem. Suppl. 1385–1391 (1982)
201. Trost, B. M.: Fundamental Research in Homogenous Catalysis, *4*, 117, Plenum Press (1984)
202. Only in the case of 2,3-di- (or higher) alkylated methylenecyclopropanes, in which the distal C—C bond is strengthened, proximal ring-opening is observed, leading to 1,3-butadienes or 1,4-pentadienes. In addition, methylenecyclopropanes with an alkyl substituent larger than methyl in the 2-position are mostly isomerized to 1,3-butadienes despite of distal cleavage reflecting the fact that elimination is faster than cycloaddition.
203. Binger, P., Bentz, P.: J. Organomet. Chem. *221*, C33 (1981)
204. Binger, P., Wedemann, P.: Tetrahedron Lett. *24*, 5847 (1983)
205. a) Jolly, P. W., Wilke, G.: The Organic Chemistry of Nickel, Vol. II, Academic Press, New York (1975)
b) Jolly, P. W.: Comprehensive Organometallic Chemistry, Vol. 8, p. 649, Pergamon Press (1982)
c) Keim, W., Behr, A., Röper, M.: ibid. p. 371
206. Binger, P., Lü, Q.-H., Wedemann, P.: Angew. Chem. *97*, 333 (1985); Angew. Chem. Int. Ed. Engl. *24*, 316 (1985)
207. Eisenberg, R., Hendriksen, D. E.: Adv. Catal. *28*, 79 (1979)
208. For an excellent review of metal catalyzed CO_2-reactions see: Sneedon, R. P. A.: Comprehensive Organometallic Chemistry, Vol. 8, p. 225, Pergamon Press (1982)
209. Hoberg, H., Oster, B. W.: Synthesis 324 (1982)
210. Hoberg, H., Burkhart, G.: Synthesis 525 (1979)
211. Inoue, Y., Itoh, Y., Hashimoto, H.: Chem. Lett. 633 (1978)
212. Ohno, K., Mitsuyasu, T., Tsuji, J.: Tetrahedron *28*, 3705 (1972)
213. a) Sasaki, Y., Inoue, Y., Hashimoto, H.: J. Chem. Soc. Chem. Commun. 605 (1976)
b) Musco, A.: J. Chem. Soc. Perkin Trans. I, 693 (1980)
214. Jolly, P. W.: Comprehensive Organometallic Chemistry, Vol. 8, 671, Pergamon Press (1982)
215. a) Maurin, R., Bertrand, M.: Bull. Soc. Chim. France, *3*, 998 (1970)
b) Isaacs, N. S., Stanbury, P.: J. Chem. Soc. Perkin Trans., 166 (1973)
216. Weintz, H. J.: Dissertation, University Kaiserslautern (1984)
217. Weintz, H. J., Binger, P.: Tetrahedron Lett. *26*, 4075 (1985)
218. Binger, P., Weintz, H. J.: Chem. Ber., *117*, 654 (1984)
219. Inoue, Y., Hibi, T., Satake, M., Hashimoto, H.: J. Chem. Soc. Chem. Commun., 982 (1979)
220. Inoue, Y., Hibi, T., Kawashima, Y., Hashimoto, H.: Chem. Lett. 1521 (1980)
221. Binger, P., Brinkman, A., McMeeking, J.: Liebigs Ann. Chem. 1065 (1977)
222. Sternberg, E., Binger, P.: unpublished results
223. Trost, B. M., Nanninga, T. N., Chan, D. M.: Organometallics *1*, 1543 (1982)

224. Büch, H. M., Schroth, G., Mynott, R., Binger, P.: J. Organomet. Chem. *247*, C63 (1983)
225. Albright, T. A.: J. Organomet. Chem. *198*, 159 (1980)
226. Gordon, D. J., Fenske, R. F., Nanninga, T. N., Trost, B. M.: J. Am. Chem. Soc. *103*, 5974 (1981)
227. Knapp, S., O'Connor, U., Nobilio, D.: Tetrahedron Lett. *21*, 4557 (1980)
228. Magnus, P., Ruagliato, D. A.: Organometallics *1*, 1243 (1982)
229. a) Piers, E., Kasunaratne, V.: J. Chem. Soc. Chem. Commun. 935 (1983)
b) Piers, E., Kasunaratne, V.: J. Chem. Soc. Chem. Commun. 959 (1983)
230. Bentz, P.: Dissertation, University Kaiserslautern 1982
231. Shortridge, R. W., Craig, R. A., Greenlee, K. W., Derfer, J. M., Boord, C. E.: J. Am. Chem. Soc. *70*, 946 (1948)
232. v. R. Schleyer, P.: Chem. Commun. 569 (1968)
233. Nametkin, N. S., Ydovin, V. M., Finkel'shtein, E. Sh., Propov, A. M., Egorow, A. V., Ivz. Akad. Nauk, SSSR, Ser. Khim. 2806 (1973); Engl. Transl. 2742 (1973)
234. a) Noyce, D. S., Fessendes, J. S.: J. Org. Chem. *24*, 715 (1959)
b) Hall jr., H. K.: Macromolecules *4*, 139 (1971)
235. Dolby, L. J., Esfandiori, S., Ellings, C. A., Marshall, K. S.: J. Org. Chem. *36*, 1277 (1971)
236. a) Kurosawa, K., Obara, M.: Bull. Chem. Soc. Jap. *39*, 52 (1966)
b) Bartlett, P. A., Green III, F. R.: J. Am. Chem. Soc. *100*, 4858 (1978)
c) Guiliano, R., Artico, M., Nacci, V.: Ann. Chim. (Roma) *51*, 491 (1961); C.A. *63*, 5066 (1965)
237. Auwers, K.: Ber. Dtsch. Chem. Ges. *26*, 364 (1893)
238. Honda, M., Hirata, K., Sukoca, H., Hatsuki, T., Yamaguchi, M.: Tetrahedron Lett. *22*, 2679 (1981)
239. a) Sakai, K. I. J., Oda, O.: Tetrahedron Lett. 3021 (1975)
b) Oda, O., Kojima, K., Sakai, K.: Tetrahedron Lett. 3709 (1975)
240. Goldsmith, D. J., Thottathil, J. K.: Tetrahedron Lett. *22*, 2447 (1981)
241. Boeckman jr., R. K., Nagely, P. C., Arthur, S. D.: J. Org. Chem. *45*, 752 (1980)
242. Trost, B. M., Renaut, P.: J. Am. Chem. Soc. *104*, 6668 (1982)
243. Sternberg, E., Binger, P.: unpublished
244. Trost, B. M., Nanninga, T. N.: J. Am. Chem. Soc. *107*, 1293 (1985)
245. Trost, B. M., Chan, D. M. T.: J. Am. Chem. Soc. *103*, 5972 (1981)
246. Kon, K., Isoe, S.: Tetrahedron Lett. *21*, 3399 (1980)
247. Baker, R., Keen, R. B.: J. Organomet. Chem. *285*, 419 (1985)

Author Index Volumes 101–135

Contents of Vols. 50–100 see Vol. 100
Author and Subject Index Vols. 26–50 see Vol. 50

The volume numbers are printed in italics

FSC
www.fsc.org
MIX
Papier | Fördert gute Waldnutzung
FSC® C083411

Zeitfracht Medien GmbH
Ferdinand-Jühlke-Straße 7
99095 Erfurt, Deutschland
produktsicherheit@kolibri360.de